Patternism

Patternism

How Patterns Define Intelligence, Consciousness, and Evolution

Thuc Nguyen

Patternism:
How Patterns Define Intelligence, Consciousness, and Evolution

By Thuc Cong Nguyen
Version 1.01 January 2025

Patternist Press
www.PatternistPress.com

Copyright © 2025 Thuc Cong Nguyen
All rights reserved.

No part of this publication may be reproduced, distributed, or transmitted in any form or by any means, including photocopying, recording, or other electronic or mechanical methods, without the prior written permission of the publisher, except in the case of brief quotations embodied in critical reviews and certain other noncommercial uses permitted by copyright law.

ISBN:
978-1-965453-00-1 (E-Book)
978-1-965453-02-5 (Paperback)
978-1-965453-01-8 (Hardcover)
978-1-965453-03-2 (Audiobook)

It is better to be stupid rather than to be blind. Stupidity can be cured with enough time and effort. Blindness can never be cured, no matter how much the object of one's desire dances before one's very own eyes.

Contents

I. A Complete Theory of Intelligence

The Patternist Approach	1
The Game	10
Universal	13
The Ladder of Complexity	16

II. A Theory of Patterns

What is a Pattern?	19
Spatial and Temporal Patterns	23
Living Patterns	25
Complexity	29
Existence	33
The Differentiation of Outcomes	37
A Semi-Consistent Universe	40
Medium Independence	42
The Paradox of the Pattern	44

III. The Search for Patterns

Adaptive Systems	53
Pattern Recognition	59
Mechanism 1. Repetition	61
Mechanism 2. Prediction	64
Mechanism 3. Natural Selection	73
Mechanism 4. Categorization	77
Universal Tools	83
The Divide	86

IV. A Theory of Theories

What is a Theory?	91
Foundational Principles	95
Constructing Reality	100
The Universal Method	107

V. A Viable Set

Merely Viable	117
Principle 1. Representation	123
Principle 2. Recognition	146
Principle 3. Reproduction	174
Principle 4. Randomization	199
A Fundamental Theory	223

Appendix

Tests of Self	237

Bibliography 251

Patternism

I. A Complete Theory of Intelligence

The Patternist Approach

The human-centric approach to intelligence is fundamentally wrong.

Although great progress has been made toward a theory of general intelligence by studying human behavior, cognition, and the human brain—the human-centric approach and the belief that consciousness is synonymous with intelligence, that only human beings are capable of displays of intelligence, is not only flawed but is the greatest obstacle to finding a complete theory of intelligence.

Intelligence is not consciousness.

The isolation of intelligence to only human beings has led to the dismissal of countless observations of other living organisms and their problem-solving abilities and solutions to maintain existence. This self-imposed limitation, based on a false premise, results in a lack of data from which the principles for a complete theory of intelligence can be drawn from.

Only by expanding the phenomena of intelligence to include all living organisms and all their activities—from reproduction to protein receptors, camouflage, and beyond—can the universal set of principles that govern all biological phenomena, principles that all living organisms use and rely on, be more readily found.

Human beings are biological.

As such, the requirement of a ***Complete Theory of Intelligence*** to be able to explain *all* biological organisms and *all* their activities must also fully explain the entirety of human-related phenomena—human intelligence, consciousness, language, science, engineering, mathematics, theory creation, technology, music, art, and so on. Anything less would be incomplete.

This absolute requirement for a *complete theory of intelligence* to be able to explain all biological phenomena is arrived at from the sheer number of observations of the results of the process of biological evolution.

Evolution is intelligent.

It is an act of hubris and human-centric thinking to claim evolution is not intelligent when it has clearly shown great displays of creativity and ability to engineer systems and mechanisms of immense complexity. After all, the process of evolution produced human beings. It produced the human brain. It produced human consciousness. It produced countless solutions to maintain existence in a universe filled with natural disasters, predation, and the decay of time—in the same universe that human beings reside in.

Evolution is stupidly intelligent.

A distinction must be made here between the theory of evolution and the process of evolution. The current modern theory of evolution, based on the synthesis of reproductive success and population genetics, is inadequate at fully explaining the process of evolution, let alone account for all the various techniques employed by biological organisms to maintain existence.

The modern theory is not detailed enough. It cannot explain or generalize the representation that occurs in genetics. It cannot explain or generalize the act of reproduction or the act of eating. It cannot explain how brains operate. It cannot explain communication and all its aspects. It cannot explain why defining species is problematic. And it does not dare go beyond the biological substrate and into the technological realm, even though it can be clearly demonstrated that technology also evolves in the same manner.

The modern theory of evolution, though powerful in its current explanatory scope, is still far from complete. Many questions remain unanswered, and even more have yet to be noticed or asked.

If anything, it is the false premise that the currently accepted theory of evolution is complete—that all biological phenomena have been solved, that there is nothing left to learn about biological organisms, that all answers can be found within the principles of

I. A Complete Theory of Intelligence

Darwinian evolution and population genetics—it is this naïve belief in what is demonstrably false that adds another layer of blindness to the existing inability of scientists and philosophers to solve the problem of intelligence and consciousness.

Had it not been for this double layer of blindness—the first false premise being that only human beings are intelligent, the second false premise being that the theory of evolution is complete—a theory capable of explaining intelligence and consciousness would have been described years, if not decades, earlier.

It is an absolute failure of those seeking a theory of intelligence to overlook the countless results of biological evolution and not use them as clues, or even outright ignore and dismiss them, in their search for a theory of intelligence.

The problem has already been solved. The process of evolution has already produced intelligent systems. The problem is not one of engineering. The task is not to gather resources in order to experiment and build an intelligent system from scratch.

No. The problem is one of description. The task is to describe what the process of evolution have already accomplish. The experiments have already been done—the data already collected. Every single organism that exists or has ever existed is an experiment, an unignorable data point. Over billions of years and countless generations, both failures and successes, the process of evolution has utilized a common set of principles to create organisms that can maintain existence in a hostile universe. In the one and *same* hostile universe that human beings reside in—and thus, by the purest of logic, must ultimately all be governed by the same set of fundamental principles.

This book will correct the massive errors caused by the human-centric approach to intelligence.

This book will completely discredit the human-centric approach by expanding intelligence to include all biological phenomena, showing that the intelligence phenomenon and biological phenomenon are one and the same. That a truly general theory of evolution that explains all biological phenomena is inseparable from a truly general theory of intelligence.

This book will complete the theory of evolution by presenting a radical new concept: *Evolution as the propagation of patterns*—living organisms as self-propagating patterns found in the environment. A new theory that will be more detailed and fundamental, fully describing the process of evolution and all its paradoxes, and will be built upon the success of the current theory in the same manner that Einstein's Relativity is built upon Newtonian Mechanics. Life itself will be redefined as a library that can find and store the recognition of patterns that helps in the propagation of that library through time and space.

This book will clearly define both intelligence and consciousness.

This book will show that intelligence came first, not consciousness, and that consciousness itself is a biological phenomenon—a product of the intelligence process. As such, a theory of consciousness can be developed by first addressing simpler biological phenomena and then progressing to more complex ones, ultimately reaching consciousness.

This book will explain the principles behind the recent advances in artificial intelligence, including neural networks, large language models, and the categorization and labeling of data for training.

This book will explain the *neuronal algorithm*, showing how neurons operate as units of pattern recognition. How neurons can effectively detect, compare, and store patterns, and thus how networks of neurons can give rise to the sought-after products of human intelligence. How memory, prediction, learning, language, theory, science, and consciousness can all ultimately be reduced to *patterns* and the interactions of *patterns*, and thus to the connections of individual neurons.

This book will reveal how all metaphors work, how all theories function, mathematics and physics operate, and how it is even possible that inspirations from biological neurons can be translated into a silicon form to produce intelligent behavior.

This book will show that there exists a common set of principles that govern all biological phenomena—including all human phenomena—and that this set of universal principles must be found and described at all costs.

All starting from first principles.

The expansion of intelligence to include all biological phenomena makes finding a set of principles for a *complete theory of intelligence* easier, not harder. There is now an overabundance of data. An overabundance of examples. An overabundance of well-documented observations. Every activity carried out by a biological organism, including the organism itself, becomes a clue—a data point—to find the set of principles. Compare this universal *Patternist approach* to the human-centric approach, which relies on a single data point—human beings—making theory construction extremely difficult simply due to the lack of available data.

The method we will use to find such a set of principles is pattern recognition. Gather large amounts of data, identify patterns within, and extract out those patterns to be used as principles for a theory. It is through this *Universal Method* that the *viable set* of principles of *Representation, Recognition, Reproduction,* and *Randomization* are found for a *complete theory of intelligence*—that given any biological or intelligent phenomenon, the phenomenon can be broken down into aspects of *Representation, Recognition, Reproduction,* and *Randomization.*

A *Viable Set* of Principles for a *Complete Theory of Intelligence*:

1. **Representation** – The use of one set of objects to represent another. For example, genes represent proteins, maps represent land, and numbers measure quantities and lengths. Language, mathematics, simulations, scientific theories, and models are all forms of representation.

2. **Recognition** – The detection, comparison, and storage of features, leading to categorization, naming, and differentiation. Examples include protein receptors, enzymes, sensory organs, self-recognition, mechanical sensors like radar or smoke alarms, and any act of identification.

3. **Reproduction** – The propagation of patterns. Beyond biological reproduction, this principle covers the manufacturing of products, the reproduction of ideas, knowledge, and skills.

4. **Randomization** – The generation of variation to search for patterns. This is seen in exploration, novelty, experimentation, meiosis, mutations, and even consciousness to carry out efficient search.

This specific set of principles is not important. It is the attempt to find a universal set of principles that underlies all biological phenomena that holds the most value. If a more inclusive set is discovered, then this particular set must be discarded in favor of the more inclusive set.

All that matters is the belief that such a universal set exists and must be found.

This is the essence of the Patternist approach to intelligence. The accounting of all biological phenomena approach is what separates Patternism from all other theories of intelligence. For a Patternist, there exists a common set of principles that all organisms—whether human beings, diatoms, or viruses, whether carried out by individuals or a population, whether occurring through culture, language, or technology—all make use of and rely on. It is through the interactions of the principles in the set that results in all biological phenomena. That all biological phenomena can be broken down into aspects of the common set of principles in the same manner that all objects in the universe, in regard to their motion, can be broken down into a single set of interacting physical laws of motion.

The question of why such a universal set of principles exists is a question about patterns. It is through patterns that Darwin and Wallace developed the theory of natural selection. It is through patterns that Newton formulated his laws of motion, and later, Einstein developed relativity. In combat sports, combatants search for patterns in their opponents to exploit. In biology, organisms are classified based on the patterns in their genetic material and morphology. The basis of all sensory organs can be generalized as tools for detecting patterns in the environment. Mathematics is considered to be the study of patterns found in strict symbol manipulations. Science seeks to record and systematize patterns found in nature. The current revolution in artificial intelligence, particularly machine learning, is due to computer implementations of pattern recognition algorithms.

The abundant use of patterns in all fields of human endeavors should indicate that there is a rich philosophical foundation for the study of patterns. Yet, as of the writing of this book, no treatise exists that specifically explains what a pattern is or how patterns are used, despite their extensive reliance by all living organisms.

Patterns have simply been ignored.

This book, in addition to completely discrediting the human-centric *view* of intelligence by expanding intelligence to include all biological phenomena, seeks to lay the foundation of *Patternism*. To clearly define patterns both practically and philosophically as well as to provide the tools of pattern recognition, describe the principal mechanisms of *Repetition*, *Prediction*, *Natural Selection*, and *Categorization* which are used by all adaptive systems to find patterns.

The belief of Patternism is the belief that patterns form a fundamental aspect of reality in the same manner that time and space are fundamental. ***Patternism is the belief that what lies at the core of all biological and intelligent phenomena—including all errors and inconsistencies—are patterns.*** All philosophical inquiries and paradoxes can be reduced to questions about patterns. Only by clearly understanding what a pattern is can any progress be made towards a theory of intelligence, towards a theory of evolution, towards a theory of language, towards a theory of consciousness, towards a theory of theory creation.

For a Patternist, all fields of study are subfields of the study of patterns.

Intelligence is no exception.

Intelligence must be viewed such that any system that can adapt and learn is intelligent regardless of whether it is conscious or not. Regardless of whether it is carried out by an individual or through a population. Regardless of whether the learning and adapting is carried out across different generations. Regardless of medium, all intelligent systems—whether biological neurons, neural networks, the immune system, genetic evolution, and even science itself—utilize a common set of principles to adapt and learn. That in the end, there is only one intelligent system—a single structure of *Universal Recognition*—that takes on many different forms.

For a Patternist, ***intelligence is the ability to find and utilize patterns***. It will be shown that only this Patternist definition of intelligence can meet the strict requirement of a *complete theory of intelligence* that is inclusive of all living organisms and all their activities.

Consciousness is no exception.

For a Patternist, ***consciousness is the ability to recognize patterns in oneself and break away from those patterns***. It is the ability to differentiate oneself from the environment, to categorize self from not-self, and to create and use representations of the self in the acts of self-referencing.

Consciousness helps an agent randomize. It helps an agent avoid getting stuck in loops. It enables an agent to learn new things by

I. A Complete Theory of Intelligence

recognizing patterns already learned and paying greater attention to novelty in order to discover new patterns.

Consciousness allows an agent to be unpredictable, especially in the presence of another intelligent agent—particularly if the other agent is human. Since human beings excel at finding patterns, and since the greatest threat to a human being is another human being, it is crucial for survival to defy following predictable patterns in case a rival is observing and seeking to do us harm.

Ultimately, consciousness is the ability to define oneself as a pattern and maintain that pattern, to understand oneself as an evolving pattern, and to direct one's own evolution. And it will be through patterns—and the *Paradox of the Pattern*—that consciousness will be explained.

It will be through patterns that language, mathematics, science, physics, music, and the act of eating can occur. It is through patterns that prediction, reasoning, perception, and deception occur.

It will be shown that it is through patterns that consciousness occurs—that consciousness is a biological phenomenon and thus the principles of *Representation*, *Recognition*, *Reproduction*, and *Randomization* can be applied to break down and analyze it. Furthermore, since consciousness involves finding of the pattern of the self—the act of turning an intelligent system onto itself—the pattern-recognition mechanisms of *Repetition*, *Prediction*, *Natural Selection*, and *Categorization* can be fully implemented to find the pattern of the self, and thus be utilized to create fully conscious systems.

It will be shown that the proposed *viable set* of principles of *Representation*, *Recognition*, *Reproduction*, and *Randomization* are merely secondary principles that can be further reduced to the fundamental principle of *Patterns*—that *Patterns* and the *Paradox of the Pattern*, form the first principle for a *complete theory of intelligence*. It will be shown that patterns lie at the core of all biological and intelligent phenomena.

And we will begin with a game.

The Game

Let's play a game.

The goal of the game is to find a set of principles that underlie all biological phenomena. That given any biological phenomenon, the phenomenon can be broken down into different aspects of the same set of interacting principles.

What is meant by all biological phenomena is any activity regularly displayed by a living organism, including the existence of the organism itself. Prions, viruses, camouflage, sexual reproduction, eating, replication, and various sensory organs are just a few examples of biological phenomena that the principles must account for.

Human beings are biological organisms. Therefore, any activity human beings engage in is, by definition, biological. Physics is a biological phenomenon because it is practiced by human beings, who are biological organisms. Mathematics, theory creation, and all forms of intellectual activity fall under this category for the same reason. Thus, the strict requirement of our game to explain all biological phenomena must also cover all human phenomena, including music, art, language, consciousness, and every other activity in which human beings participate. This extends beyond activities driven solely by genetics and includes engineering, manufacturing, technology, science, and even deception. Computers, smartphones, cars, and various forms of entertainment are all included.

Why include technological and industrial phenomena as biological phenomena?

The argument that computers, smart phones, cars, and processes like the scientific method, engineering design, and manufacturing assembly are not part of the biological phenomena is wrong. While such electronic goods and the processes involved in their production are not biological in themselves, they are still the results of human activity and so must be categorized as part of the biological phenomena. After all, are computers, smartphones, and cars not products of human activity? Is it not true that science, physics, and

I. A Complete Theory of Intelligence

mathematics are conducted by human beings, making them part of the biological phenomenon? To claim that these activities are not biological is akin to saying that bird nests, termite mounds, and spider webs are not biological phenomena.

In essence, the goal of our game is to identify a set of principles in which together form a *complete theory of intelligence*. A complete theory that seeks to account for all intelligent behavior and phenomena, not just those related to human beings.

Why call it a theory of intelligence?

If we are to hold the word *"intelligence"* as the pinnacle of complexity—in terms of inventiveness, practicality, and creativity—then there is no other name we can give a theory that seeks to explain all biological phenomena, including all human behavior and activities.

The evolutionary process is an intelligent process.

We must reject the notion that consciousness is synonymous with intelligence. Human beings, human intellect, and human consciousness are products of biological evolution—an undeniable fact. The immense engineering and creativity displayed by evolution force us to recognize that the process of evolution utilizes certain fundamental principles inherent to the universe to solve the problem of existence in a world full of predators, natural disasters, and decay brought on by time—a feat evolution has managed for over a billion years.

The argument that evolution is not intelligent because it is a long, slow, and random process is an act of hubris and human-centric thinking. If anything, it is the opposite. Evolution's ability to sustain a process over such an immense timespan is a testament to its ingenuity and effectiveness. What other process can be named that is capable of creating, spreading, and enduring through the unpredictable violence of our universe for over a billion years?

What does it matter that the process relies on randomness and unfolds over immense timescales? What does it matter how long it takes, or the method used if the problem is solved in the end? If anything, it should be celebrated—the difficult work has already been done. The fundamental principles have already been created and refined through a billion-year process.

The fundamental principles are already in use. Now, it is a matter of describing them.

What are those fundamental principles?

And why have scientists and philosophers not sought to identify and describe such a foundational set?

When searching for a complete theory of motion, do we not aim to find a set of universal principles that underlie all phenomena related to motion?

When developing a quantum theory, do we not seek to uncover universal principles that explain all phenomena related to quantum mechanics?

When working toward a comprehensive theory of electromagnetism, do we not strive to identify the set of universal principles that govern all electromagnetic phenomena?

Do all biological phenomena not exist within the same universe? Are human beings not biological? If so, why has no similar effort been made to identify a set of principles that underlie all biological phenomena—a set that would naturally cover all human phenomena as well?

The human-centric approach to intelligence is fundamentally wrong.

It is an absolute error to not recognize that both human beings and evolution rely on the same set of universal principles to solve problems and to maintain existence.

Patternism will correct this error.

Patternism is the belief that such a set of fundamental principles, utilized by the process of evolution, exists—and that these principles together form a *complete theory of intelligence.*

I. A Complete Theory of Intelligence

Universal

Why should such a set of universal principles exist?

Why should we assume that the principles governing human intelligence, consciousness, language, and self-recognition are the same principles that govern sensory organs, protein receptors, and genetics? Why should we assume that the principles governing all biological phenomena are universal?

This same question can be asked about the laws of motion in physics. Why do we assume that all objects in this universe, in regard to their motion, obey a single set of physical laws without exception? Why should the principles governing the motion of a racecar on Earth be the same as those governing the orbit of Pluto or a binary star system a hundred light-years away? Why do we assume the laws of physics are universal?

Historically, physicists assumed a single, universal set of physical laws was due to experimentation. Scientists working with weights, springs, pendulums, and similar objects observed patterns in their behavior, eventually formalizing these patterns into a theory capable of making predictions, such as determining where a cannonball will land given certain initial conditions. This theory of motion was then applied to various objects and found to be predictively accurate through continued experimentation. From thermodynamics to the movement of celestial bodies to the refraction of light through lenses, the laws of motion have proven generally applicable across many domains of study, explaining such a wide range of physical phenomena that they are assumed to be universal.

The same can be said about all biological phenomena. By understanding precisely what a pattern is, it will be shown that a universal set of principles exists to be found. This does not imply that there are no intermediate theories or principles governing only human intelligence or only genetic evolution. A theory of human intelligence may be distinct from a theory of genetic evolution, but the assumption of universality implies that a deeper, more fundamental set of principles governs both, as well as all other biological phenomena—including science and manufacturing. This set of principles, which has not yet been found or described, must be found and described, for it is the very foundation upon which all else is built.

Philosophically, we can argue for the universality of such a theory based on the fact that all phenomena exist within a single physical reality. If all biological phenomena exist within the same physical reality and are fundamentally governed by the same set of physical laws applied uniformly across all, it follows that all biological phenomena would also be governed by a shared set of biological principles. The alternative would be a universe divided into partitions, where the ultimate laws of reality differ in various places and apply unequally.

But we do not need to go that far to argue for the universality of a *complete theory of intelligence*. As far as we are concerned, all the laws of physics apply equally on Earth. And as far as we have observed, all biological phenomena have been found on Earth. We can argue that, at the very least, universality is locally true.

We do not need to accept the argument of a single universe with universal laws to acknowledge the existence of a single set of principles governing both biological and intelligent phenomena. We can argue for universality based on practicality.

There are two major practical reasons to accept and proceed as though universality is true.

The first reason is simplicity. Denying universality would force us to create multiple theories to explain different phenomena. Cannonballs fired in Europe would have different trajectories than those fired in America. There would be different theories of motion in physics. There would be a theory of genetics that applies only to genetics, a theory of neurons that applies only to neurons, and a theory of camouflage that applies only to specific species. Not only would each theory be distinct, but there would also be no shared commonality or overlap—no general solutions linking phenomena to one another.

And why not divide it even further? Why not have a theory that applies only to one location at one specific time? A theory that works in New York would not work in Houston, and a theory that applies in 2020 would not work in 2025. Denying universality leads to a proliferation of theories we must memorize and constantly keep track of.

The second reason is engineering. We want to build and create with our theory. Denying universality would render all engineering pointless. There would be no reason to expect the wings of birds and the wings of airplanes to work in a similar manner. We would expect no similarities between the fins of penguins, sharks, and whales—or even between two individuals of the same species. We would not be inspired to build camera systems by observing how biological eyes work. No biomimicry in robotics. Nor would there be any point in pursuing artificial intelligence or reproducing human intelligence, because if we deny universality, there is no reason to expect systems that work in a biological substrate to be transferable to silicon transistors. No reason to believe evolutionary algorithms would work. No reason to expect neural networks to function.

But the fact is that it does work. The wings of birds operate in a similar manner to the wings of airplanes. The fins of penguins, sharks, and whales share similar shapes. Recent advances in artificial intelligence—especially in machine learning and the development of neural networks—were inspired by human neurons. Given the existence of so many general solutions transferable across diverse mediums, we expect there to be a universal set of underlying principles.

A set of universal, underlying principles that must be found and described at all costs.

The Ladder of Complexity

Although it sounds counterintuitive, the bottom-up Patternist approach of explaining all biological phenomena on the way to explaining intelligence is much easier to accomplish than a top-down, human-centric approach that focuses solely on studying human beings.

By accepting universality, we now have a starting point for finding a theory that can explain human intelligence and consciousness by first addressing simpler biological phenomena. If we think of all biological phenomena as steps on a ladder—with complex phenomena, such as human intelligence and consciousness, at the top of the ladder and simpler phenomena, such as viruses, protein receptors, and the act of eating, repairing, and reproduction, at the bottom—we now have a clear path toward zeroing in on a theory that can explain human intelligence in addition to all other biological phenomena.

We start at the bottom of the ladder and craft a theory that explains simpler biological phenomena, which has fewer variables to account for. Once we establish a theory that satisfies our requirements, we can proceed up the ladder, tackling increasing levels of complexity and further testing our theory as we go. If our theory reaches a dead end—failing to explain a particular biological phenomenon—we can step back, reevaluate, and modify it, knowing we may have overlooked a critical factor. By correcting and refining our theory before moving forward, we can climb *The Ladder of Complexity* with increasing confidence in our theory's accuracy.

The alternative to this bottom-up Patternist approach—the top-down, human-centric approach—is far more difficult. Starting with human beings and human intelligence leaves us with no way of knowing how far down we must go to identify the necessary components for building a working set of principles. There are too many variables and extraneous factors we have yet to encounter, account for, or explain.

We do not know what is important and what is not.

I. A Complete Theory of Intelligence

Can we find a theory of human intelligence by studying only human behavior and language? What about the environment in which the human body and language exist? What about the human brain? Does the brain exist in isolation? What about neurons and their connections? Don't we need a theory of protein receptors, given that neurons rely on a variety of them, not to mention neurotransmitters? Wouldn't these and their interactions play a crucial role in a theory of human intelligence? Neurons themselves contain genetic material and consume nutrients. And what about the mitochondria within neurons? Given that mitochondria have their own set of genes, is this not important?

The complexity of the top-down approach increases, leaving us in a quagmire with no clear path forward and no way to know if we're even on the right track. The top-down, human-centric approach proves to be far more challenging than the bottom-up Patternist approach.

So, we must climb the ladder.

To find a *complete theory of intelligence,* we must start at the very bottom of *The Ladder of Complexity* and steadily work our way up. No steps are to be left untouched. No biological phenomena can be ignored. All data points must be connected. Accounting for all biological phenomena is what separates Patternism from all other theories of intelligence and makes it superior.

Patternism—the belief that patterns lie at the core of all biological phenomena—will be built on solid ground. We will not stray from reality. We will begin with reproduction, the acts of eating and repairing, at genetics, and protein receptors. We will work our way up to sensory organs, multicellularity, the immune system, and sexual recombination. We will explain how symbols written in ink on paper can make accurate predictions and reflect reality, thus explaining how all forms of representation—language, mathematics, physics, science, and theories—operate. We will arrive at engineering, neural networks, language, and even the very act of theory creation itself.

We will account for all biological phenomena and all human phenomena, especially those that are well-documented, widely observed, and have practical applications.

We will steadily climb our way up.

Each step we take will follow coherently from the previous. And by the end of this book, we will reach human consciousness. But make no mistake—the *human-centric approach* to intelligence, the belief that a theory of intelligence can be found by studying human beings alone, and the *human-centric view* of intelligence, the belief that only human beings are intelligent, and that consciousness is the same as intelligence, must be completely discarded. Everything on our ladder is the result of an intelligent process. Human consciousness may sit at the top of the ladder, but it is still a product of the intelligent phenomenon.

II. A Theory of Patterns

What is a Pattern?

A Pattern is a discernible regularity.

All living organisms make use of patterns. The human immune system identifies patterns in pathogens to initiate an appropriate response. Sensory organs function as tools to detect patterns in the environment. A dictionary catalogs the patterns that a language and its users recognize. We find patterns in music and poetry. Mathematics is the study of patterns. Science rigorously documents patterns found in nature.

If patterns are so essential across so many different human endeavors and to life itself, why is there no comprehensive philosophical treaty exploring what exactly a pattern is—how patterns are searched for, stored, and used?

Why is there no general theory of patterns?

Is it because the word "*pattern*" has been reduced to surface-level concepts, like textile and design prints, to mere decorations? Or is it because patterns are so universal that they are hard to notice without contrasting them to a universe where they are absent—like asking a deep-sea fish, "What is water?"

Yet, in all discussions of intelligence, patterns are frequently mentioned without any in-depth examination of what exactly a pattern is.

Patternism is the belief that patterns form the first principles for a *complete theory of intelligence*. All biological and intelligent phenomena can be reduced to questions about patterns. Without a theory of patterns, there can be no theory of intelligence. Any attempt to understand intelligence must begin with a fundamental understanding of patterns.

All else will follow.

A Patternist definition of a pattern must be universal. It must cover not only everyday uses of the word *"pattern"* but also applications across technical fields. This includes textile patterns, design prints, number sequences in mathematics, how a martial artist studies an opponent's patterns, how patterns are identified in machine learning, how neurons detect and store patterns, and even the phenomenon of reproduction. A Patternist definition must be broad enough to cover all these examples.

So, what is a pattern?

A Pattern is a discernible regularity.

While there is complexity in what it means to be *"discernible"* and *"regular,"* this simple definition allows us to cover a wide range of phenomena. Reproduction, perception, prediction, deception, reasoning, and representation all rely on this basic concept of patterns.

A pattern is made up of features.

A feature is anything that can be used to separate two entities in a consistent manner. For example, color is a feature that allows us to differentiate ripe fruit from unripe fruit. Hair and mammary glands are features used to separate mammals from other organisms. Negative charge is a feature of electrons. Inertness is a feature of noble gases. Terms like *"properties," "descriptions,"* and *"attributes"* are all similar to how we define a *feature*. Anything can be considered a feature as long as it maintains consistency in separating entities.

Once the separation of entities occurs and features are defined, it immediately follows that features can be used to group entities together in a consistent manner. For example, we can group together all entities that share the feature of being the color blue. We can group all entities that have a specific gene. We can group all entities that have the features of having hair and producing milk. In effect, ***a feature is anything that can be used to both separate and group entities in a consistent manner***.

A pattern can be viewed as a collection of features. If enough features are detected, the recognition of the pattern is triggered. We can define a pattern as *strict* if recognition is triggered only when all features are detected. Conversely, a pattern is *fuzzy* when only some features need to be detected for recognition. Each feature can also have different weights in triggering recognition, and some may even have negative weights that inhibit recognition.

Features themselves can be viewed as patterns if they maintain consistency in the separation of entities. From this interplay, recursive and hierarchical structures can emerge. Features can consist of sub-features, and multiple features can combine to form a larger, overarching feature. Similarly, patterns can combine to form super-patterns or be broken down into sub-patterns.

The key distinction between features and patterns is that a pattern is any discernible regularity, whereas a feature is defined by its ability to separate and group entities in a consistent manner. This separation and grouping of entities—essentially the formation of sets, sub-sets, and super-sets—plays a fundamental role in many different phenomena.

With patterns and features defined, we can now look at number sequences to illustrate how patterns function in mathematics.

- 2, 4, 6, 8, 10, 12... follows a pattern where the defining feature is that all number are even.

- 5, 10, 15, 20, 25... forms a sequence with the feature that each number is divisible by 5.

- 2, 3, 5, 7, 11, 13... consists of prime numbers, characterized by the feature that their only factors are 1 and themselves.

Given a set of numbers, we can define a feature shared by all members of the set. Knowing the features of a set allows us to predict and generate its members, as they follow a specific pattern. All mathematical formulas and equations work in the same manner. Every equation can be viewed as a collection of variables—a list of features—and their relationships to each other. If we view mathematics purely as symbolic manipulation, then the results of a series of manipulations will be strictly regular and consistent, making them patterns in themselves—provided the system of allowed manipulations remains consistent at each step.

Textile prints and pattern design work in the same manner as number sequences. Once we identify the specific features of a design—such as the colors, shapes, and their arrangement—those features will repeat in space. As a result, if we know the features of a pattern design, we can predict and even paint and draw in the design that will appear in a given blank area of space farther away.

II. A Theory of Patterns

Spatial and Temporal Patterns

Patterns are often recognized visually, as in geometric shapes, designs, and decorations where the features of a pattern repeat spatially.

We begin with basic regular polygons: triangles, squares, pentagons, hexagons, and so on. The pattern these shapes follow is that all angles within the shape are equal, and all sides have the same length. By adhering to this pattern, we can generate any regular polygon, whether it has ten sides or a hundred. In essence, we have an algorithm to generate regular polygons.

The same principle applies to more complex structures. Objects like pyramids, cylinders, double helices, and obelisks—anything with symmetry, whether rotational, reflective, translational, helical, or even fractal—contain embedded patterns. By observing only a part of these structures, we can use their symmetry to deduce the whole and generate the object algorithmically.

All algorithms generate patterns. The reverse is also true: if we notice a pattern—a repetition in features—then there is an underlying algorithm that we may not yet fully understand. The task is to use the repetition in features as clues to discover the underlying algorithm. All science employs this methodology, focusing on regularities to create theories. Theories of evolution, motion, electromagnetics, and matter were all developed when scientists observed patterns in nature and sought to understand the underlying algorithm—the key features and principles governing these patterns. We will use this same approach to derive the algorithm, the set of principles for intelligence and all biological phenomena.

Patterns do not only appear spatially; they also occur in time, with features repeating temporally.

Music is a common example of temporal patterns. There are regular intervals between notes that result in beat, tempo, and rhythm. If we fail to detect these repetitions, music ceases to be music and instead becomes noise. This applies to rhyming in poetry, where the last syllable of one verse repeats at the end of the next. This pattern is evident in poetic forms, whether in Homer's dactylic hexameter or the 5-7-5 syllable structure of Haiku.

Music's patterns can be complex and evolve over time. Each instrument may follow its own pattern, while together they create intricate combinations, with transitions connecting one pattern to another.

Repetition in events also counts as a temporal pattern. The changing of the seasons follows a cycle: spring leads to summer, then fall, then winter, and the sequence repeats. Celestial movements follow regular cycles, marking time for societies. Holidays like New Year, Thanksgiving, and Christmas are recurring events, each with associated themes and symbols.

Detecting and recording temporal patterns requires a suitable medium to translate them into spatial representations. For example, written music translates auditory patterns into a visual format that can be stored and referenced in a book. Similarly, a calendar records notable events over time. Through repeated observations, the calendar reveals patterns that enable predictions about future events, such as anticipating heavy rains or droughts, or preparing for colder temperatures and ensuring sufficient food supplies to last through the winter. In this way, temporal patterns become tools for planning and survival.

A system that stores and retrieves knowledge of temporal patterns is called memory.

Neurons in the human brain and DNA in living organisms are examples of memory systems, each optimized for different timescales. Patterns unfolding over seconds to several decades are detected and stored in the human brain through networks of neurons. In contrast, patterns that occur over tens of thousands of years are stored in DNA through evolution. Patterns that span a hundred to several thousand years—too long for individual memory and too short for DNA—are stored externally in a society's culture, through stories, books, language, traditions, institutions, and religion.

Living Patterns

A pattern is a discernible regularity.

Anything repeatedly found sharing the same features—the same color, form, or behavior—is considered a pattern. It directly follows that living organisms are patterns found in the environment.

This is undeniable. Biological organisms are not just informational patterns; they are literal patterns—recurring entities, structures, and forms—found in the environment. Ants are a pattern, with repetition in body form easily observed in a trail of ants. Cockroaches repeatedly appear in swarms when the lights are turned on in a sticky kitchen, even after many were killed the day before. Specific trees, with shared features like leaves, flowers, and fruits, appear repeatedly throughout a forest, as do blades of grass. It is undeniable that living organisms are regularities found in the environment.

Even the way scientists study living organisms shows that the field of biology, from its very beginning, has been the study of living patterns. If this were not true, what would be the point of classifying species in taxonomy? Why group and catalog organisms based on their similarities and differences in features? And in modern evolutionary biology—examining organisms in terms of population genetics—do biologists not classify organisms based on patterns, on the similarities, differences, and repetition found in their DNA?

Living organisms are self-propagating patterns found in the environment.

The current view of evolution, which treats organisms in terms of their reproductive success and population genetics, must be updated to a more fundamental view: evolution as the propagation of patterns through time and space. This Patternist view of evolution shifts the focus to understanding what it means to self-propagate and how patterns change over time in a world subject to decay, natural disasters, and competition from other self-propagating patterns.

A radical shift occurs when organisms are viewed as self-propagating patterns. For one, it eliminates the need to treat reproduction as a fundamental principle of evolution. Instead, patterns become the first principle, with reproduction deriving from patterns through the *Paradox of the Pattern*.

Second, patterns are composed of features, which act as components of an algorithm that generates the original pattern. Applied to living organisms, this approach integrates genetics and reproduction into a unified framework rather than treating them as separate principles. Reproductive success and genetic representation are now directly derived from patterns, rendering the modern synthesis of evolution unnecessary.

In molecular biology, an organism's physical traits can be traced back to DNA sequences. These DNA sequences operate like an algorithm—the genetic code that links DNA to proteins, leading to the reproduction and development of the organism.

The algorithm that transforms a sequence of DNA into a living organism is highly regular and adheres to well-defined patterns. DNA is organized into sets of three nucleotides called codons, each coding for a specific amino acid. These amino acids form chains that fold into proteins, ultimately giving rise to an organism's traits. This process of representation is only possible when patterns are understood at a fundamental level.

Lastly, by viewing organisms in terms of patterns, reproduction can be generalized beyond biology. For a Patternist, the act of reproduction must be expanded to cover a broader array of phenomena, not limited to biological organisms alone. What is missing is a general theory of reproduction—a theory that would include everything from simple replicators like prions, viruses, and bacteria, to mitosis in eukaryotes, to sexual reproduction, to multicellular organisms, and the reproductive strategies of ant and bee colonies. It would also extend beyond biology to manufacturing, covering everything from simple tools like arrowheads to assembly lines, blueprints, and complex objects such as cars, smartphones, and computers.

The argument that manufactured goods should be separated from biology, and that there are no similarities between biological processes

II. A Theory of Patterns

and manufacturing, is wrong. Manufactured goods are regularities—patterns found in the environment. There are even similarities between a blueprint schematic and DNA, as both serve as representations that can be used to regenerate the original object.

When we consider the entire process of manufacturing, we see that goods evolve over time just like living organisms. If we look at the history of computers, we find that computers have evolved and achieved greater complexity over time. The process of evolution is mirrored. A smartphone brand has a lineage traceable to a single prototype. We observe the development of variations, competition among brands, and the extinction of certain computers and manufacturing companies. The process of evolution ceases to be limited to mere biology and is applicable to technology, generalizing evolution across all domains—even extending to the reproduction of ideas and knowledge through the phenomena of books, libraries, media, and universities.

Such similarities exist, and for a Patternist, they are to be expected. Through patterns, we connect mathematics and algorithms to biology, and biology to manufacturing. The entire field of biology is no longer isolated. There is no longer a "*magical wall*" separating life from the rest of the universe. Direct transitions can now be made between inorganic matter and organic matter. The processes, steps, and forces governing this universe that give rise to living organisms are no longer beyond reach.

All from understanding what a pattern is.

The study of life is the study of living patterns. Living organisms are patterns found within the environment. But we will not stop here. We will go further and define life as anything that can find and make use of patterns. Anything capable of recognizing and utilizing patterns is alive and will display life-like behaviors and properties.

For a Patternist, intelligence is the search for and utilization of patterns. Thus, the study of life is the study of intelligence.

Is this not true?

Do we not study human beings—a living organism—as part of the phenomena of intelligence? Did life not produce human beings, the human brain, and human consciousness? Has life not adapted, learned, and evolved—activities that are part of intelligent phenomena and hold immense utility? Do we not treat anything that is alive or possesses life-like properties as either a valuable ally or a dangerous enemy that can kill us? And does life—anything living that we once thought we controlled—not eventually evolve beyond our control and even surprise us?

Life, then, is intelligent. Anything alive will display acts of intelligence. Anything intelligent will display life-like behavior.

All life is intelligent. There is no separation between intelligent life and life. Labeling living organisms as either *"intelligent"* or *"non-intelligent"* is a human-centric error—an act of arrogance and hubris. Anything that can adapt, learn, and evolve is intelligent. Anything that can find and make use of patterns is intelligent.

All life is intelligent.

The biological phenomena and the intelligent phenomena are one and the same.

II. A Theory of Patterns

Complexity

Patterns can evolve over time, resulting in outcomes that are radically different from their origins.

In evolution, an organism may appear identical from one generation to the next, but over millions of years and billions of generations, it can change dramatically, eventually sharing no visual features with its earliest ancestors. Through this process, every human being can be traced back to a reptile-like, furry rodent, then to a fish, and ultimately to a single-celled bacterium.

This process of evolution—the transformation of patterns over time—is not limited to biological life but can also be generalized to technological development and cultural evolution. Firearms, airplanes, smartphones, automobiles, computers, language, and institutions—whether military or educational—have all evolved and reached greater levels of complexity over time.

Evolution is a universal process. All technological innovations operate under an evolutionary process. The same forces that lead to biological complexity—natural selection, competition, random mutation, and viral infections—have comparable analogs in the technological and cultural domains. For example, in the storage, duplication, sharing, and creation of information, the phenomenon of sexual reproduction can be compared to the invention of the printing press and the internet. Soon after the emergence of each of these phenomena—sexual reproduction, the printing press, and the internet—a revolution in complexity follows.

What is the alternative? What happens if we continue to deny the universality of evolution and limit it to biology alone?

The answer is simple.

We will remain ignorant of the universal forces that drive technological innovation. We will be uninspired. We will continue to see biology as isolated, even though evolution has produced organisms that can fly, engineer, and think.

We will remain blind.

Nothing occurs in a vacuum, and technological innovation is no exception. The forces that enable evolution to produce the human brain and human beings are the same forces that drive technological innovation.

There is no isolation. Scientists and engineers do not work alone. Institutions and infrastructures facilitate the communication of information from one scientist or engineer to another, allowing progress to be copied, combined, experimented with, and improved upon. Similarly, individual organisms do not operate in isolation. Through sexual reproduction and genetics, traits are shared, tested, and subjected to natural selection within populations, enabling species to evolve and reach greater levels of complexity.

Complexity to what ends? We know that a Tesla electric vehicle is more complex than the Ford Model T. We know that a slim LED TV is far more complex than a bulky cathode-ray tube. We observe that, in any technology, the modern version is more complex than its predecessor. Given two entities of the same lineage, we can consistently determine which is more complex. But what exactly does it mean for an entity to be more complex? More specifically, what does it mean for a pattern to increase in complexity? A universal definition is needed.

For a Patternist, **complexity is the quantity of patterns utilized by an entity**. The more patterns an entity possesses, uses, and exploits, the more complex the entity comparatively becomes.

Patterns build upon patterns. Complexity builds upon complexity. Each solution to a problem becomes a new pattern ready to be exploited, sparking further adaptation and innovation in an endless cycle.

The venom of a scorpion becomes more potent. A cheetah grows faster. Tanks and airplanes become deadlier. Bombs become more destructive.

There is no isolation. The environment will always contain a multitude of intelligent systems. Intelligence is the ability to find and exploit patterns. When intelligent systems compete, they drive each other toward ever-higher levels of complexity in a perpetual arms race.

II. A Theory of Patterns

Intelligent systems exploit, challenge, and adapt to one another. One system attacks; another builds defenses, prompting attackers to refine their strategies while defenders enhance their resistance. Each advantage compels the other to innovate, counter, and seek the upper hand. This dynamic accelerates both evolution and innovation, deepening each system's complexity as they push each other toward new solutions and adaptations.

Through this relentless interplay, intelligent systems—whether biological or technological—fuel ever-greater complexity. Each iteration builds on the last, layering solutions upon solutions, patterns upon patterns.

The direction of technological evolution is undeniably toward increasing complexity, as is biological evolution. A lineage of biological phenomena can be traced by increasing complexity: from prokaryotes to eukaryotes, to sexual reproduction, to multicellular life, to organisms with nervous systems, to adaptive brains, to human beings, and beyond to science, engineering, and artificial intelligence.

Although the basic forces underlying evolution that generate variation may be random, the selective pressures of merely maintaining existence in a physically demanding universe drive evolution to develop patterns of greater complexity. Unfortunately, current theories of evolution cannot fully explain the increasing complexity of biological life, let alone connect it to the innovation observed in technology.

Patternism provides an answer. Patternism not only explains evolution's trend toward greater complexity from first principles but also predicts its ultimate goal.

From our understanding of patterns, we see that reproduction is the act of self-propagating patterns. To adapt is to make use of patterns found in the environment. To predict, to deceive, and to eat are all acts of pattern utilization. Given the utility of patterns, it follows that all organisms must find and use patterns to survive in a demanding universe.

Thus, organisms must develop the ability to find and utilize patterns. Evolution's goal, therefore, is to develop faster and more efficient ways to find and use patterns. Genes store patterns across generations. Sexual reproduction enhances this by allowing patterns to be shared and combined within populations. The development of the nervous system and the storage of patterns within neurons allow patterns to be found and utilized within a single lifetime. The invention of language allows patterns to be stored outside the individual and within culture. The invention of writing and the printing press accelerated this, as did the internet, which has expanded the speed and breadth of pattern sharing yet again. The current pursuit of artificial intelligence aims to create systems that find, recognize, and store patterns, allowing these systems to be mass-produced and put to work without complaint.

For a Patternist, the ability to find and utilize patterns is intelligence. As a result, evolution seeks to create faster and greater levels of intelligence. It must be emphasized that, although Patternism describe evolution as having goals, the drive toward greater intelligence is not due to any intentional agency within evolution but rather to the natural process of elimination: among all variations generated, those that help identify and exploit patterns will outlast and out-compete those that do not.

Human beings did not consciously decide to understand intelligence. The belief that we chose to understand intelligence, find a theory, and replicate it in machines is a lie. Our pursuit of understanding intelligence, of knowing what we are, has always been driven by the unconscious forces of a universe that is dominated by the utility of patterns.

We human beings are merely the result of—and just a step in—the process of patterns searching for patterns.

Existence

Why do patterns exist? Why should the universe contain any regularities at all? From our observations and engineering, we know that the laws of physics—gravity, electromagnetism, and quantum mechanics—are regular and universal, but why should this be? We know from experimentation that all electrons are the same, all protons are the same, and all neutrons are the same, but why should any two subatomic particles share the same properties and features? Why is it that we can generalize, categorize, or name anything at all?

For a Patternist, there are no answers to *"why."* For a Patternist, patterns are an inherent property of the universe in the same way that space, time, and matter are inherent properties. Asking why the universe contains such regularities and allows them to form is difficult, if not impossible, to answer. Instead, Patternists focus on what can be built from these regularities.

Starting with regularities and aiming to lay the philosophical foundation of Patternism, we begin by defining existence in terms of patterns, with the intent to account for all biological phenomena in a practical framework.

Exist as what?

For anything to exist, it must have a form—a set of features that display regularity—and be able to maintain that form for some time. For example, a species is defined by genetic patterns consistently found within its population. A nation is defined by its geography, values, language, dress, and institutions. A border is marked by physical barriers. The President of the *United States* has a set of features, like being elected and residing in *The White House*. Nitrogen has a specific number of protons, atomic weight, and characteristic reactions. An individual has a name and defining attributes like height and eye color on a government-issued ID.

To exist means to exist as a pattern—as a set of regularities.

Object Permanence refers to the regularity with which an entity exists in time and space. Both time (moment-to-moment) and space (position/location) are features that can be used to describe an entity. Regular in time and regular in space. Continuous in time and continuous in space. Objects do not simply appear, vanish, and then reappear again. They don't teleport around. They evolve through time and space. When we park our car, we expect our car to remain in the same spot when we return. If our car disappears, we are alarmed. The same reaction occurs when we watch a magic trick, such as a magician making a coin disappear in one hand and reappear in the other. It captures our attention and surprises us when our sense of object permanence is violated. If objects teleport or vanish suddenly, we feel something is wrong, that a trick is being played, and we instinctively search for the cause of the violation, especially if it occurs within our sight.

Object permanence must be viewed in terms of patterns. By understanding permanence as patterns, we can use the tools of pattern recognition—*Prediction*, *Repetition*, *Natural Selection*, and *Categorization*—to observe and describe the permanence of an entity as well as the features of its existence.

Conversely, **Object Impermanence** is the recognition that entities do change, evolve, or disappear over long periods of time. It is the recognition that nothing remains constant forever, especially over extended intervals.

The balance between permanence and impermanence drives the need to learn. Intelligence is active, not static. For a Patternist, intelligence is the ability to find and exploit patterns. In the divide between object permanence and impermanence lies the necessity for intelligence: to learn the extent to which an entity exists, to discard what is no longer useful, and to learn again.

Things change and evolve over time. If we identify a pattern, we must constantly update the pattern. We must remain aware and adaptable, knowing that things are regular but only to a certain degree and for a limited time. Zebras must learn the seasonal permanence of watering holes. Ants must learn about the location of food, which then disappears, requiring the ants to search again. Bees must find flower patches, communicate their locations, and then move on once resources are depleted.

Within this balance lies the complexity of existence. How consistent is an entity? Is it regular enough to take advantage of? Where does it exist, and for how long? What are the features of a pattern, and to what degree is it regular? It is within this balance between permanence and impermanence that the need arises to find patterns and use them before they disappear.

Exist relative to what?

The extent and duration of an entity's existence can range from objective to subjective. The objective view is evolutionary, where regularity and duration are shaped solely by natural selection. In contrast, the subjective view depends on another entity, such as an agent, which is defined as anything that can act.

Our concern is how an agent detects and perceives the existence of objects. In the most fundamental sense, an entity exists relative to an agent—whether enzyme, bacterium, virus, dog, or human—if the agent can use the entity to differentiate its actions in a consistent manner. For example, a host cell exists relative to a virus because the virus can detect the host cell using proteins on its capsid shell, prompting the virus to inject its genetic material. A dog can detect scents that humans cannot. Astronomers did not know that the planet Pluto existed until disturbances in Neptune's orbit were observed.

Entities exist for an agent if they can be sensed and detected. The more regular and predictable an entity is, the more confident an agent becomes in its existence, even if indirectly perceived through representations or instruments.

For a Patternist, all that exists are patterns. Questions like *"What is real?"* or *"What is not real?"* are irrelevant. The only question that matters is *"How regular is it?"* Is it regular enough to be reproducible for engineering and thus be called a solution? Is it regular enough to serve as a principle for a theory? Is it regular enough to create representations and make predictions? Is it regular enough for an individual or a population to rely on for survival?

Although both views—a separate objective reality and a relative subjective reality—can be argued for from a Patternist perspective, our practical framework assumes an objective reality where patterns exist. However, in the process of identifying and storing these patterns, a subjective reality is formed and inhabited. In other words, there is an external universe containing various patterns. A subset of these patterns is recognized and stored by an agent using representations. The representations then become the agent's subjective reality—an internal, man-made reality that can be inaccurate and prone to errors.

In this framework, the human brain serves as the filter between external reality and the internal reality of perception. Thus, all we truly have access to are representations of reality. When our representations are accurate and align well with external reality, we can navigate the world effectively and make accurate predictions. When our representations are wrong, we stumble, make errors, and experience cognitive dissonance. We encounter illusions, hallucinations, bias, and other distortions.

The Patternist view of existence as a set of patterns plays a central role in defining the self. The use of the word "*I*" is a representation of the self through language. Anything that can be represented, categorized, labeled, and used consistently to a practical degree is a pattern. Therefore, "*I*," the self, is a pattern.

For a Patternist, the effective use of the word "*I*" proves self-existence. But this does not mean that only agents capable of using language to self-identify have a sense of self. Instead, the self exists on a spectrum of increasing complexity, from self-differentiation to accurate self-description and the ability to predict and control one's actions.

In line with Descartes' "*Cogito, ergo sum*," for a Patternist: "*'I' am a pattern. I am an evolving pattern. I choose what I evolve into. Therefore, I am free.*"

II. A Theory of Patterns

The Differentiation of Outcomes

We live in a universe where things are different, and they are different to a regular degree.

And we live in a universe where the interactions of different things lead to different outcomes with regularity.

This regularity in differential outcomes is what we will describe as *"The Differentiation of Outcomes,"* where the consistent differences between entities, including the environment itself, are substantial enough that their interactions lead to predictably different results. Conversely, interactions of the same entities under the same conditions will yield the same outcomes.

In short, we live in a universe that permits cause and effect.

The *Differentiation of Outcomes* occurs throughout physics. Photons emitted by helium gas have consistently different wavelengths than photons emitted by argon gas because helium and argon are consistently different from each other. The interactions of oxygen atoms with hydrogen differ from those of oxygen with iron, and so on. Electrons, protons, and neutrons are distinct from each other, and these differences are highly consistent. When electrons, protons, and neutrons interact, variations in their interactions—such as the number of protons in the nucleus—result in the regularities we describe as chemical elements.

The *Differentiation of Outcomes* is present in all describable interactions in the universe. It is through the *Differentiation of Outcomes* that variations in genetic sequences lead to differences in reproductive success, allowing natural selection to occur. It is through The *Differentiation of Outcomes* that accurate predictions can be made, sensory organs such as eyes, ears, and nose function, and categorization occurs, allowing entities to be grouped and separated into sets. It is through the *Differentiation of Outcomes* that features are defined. It is through *Differentiation of Outcomes* that science and physics are carried out, allowing the universe to be studied at all.

The ***Differentiation of Outcomes*** has three components:

1. **Consistent Differentiation of Entities** – The differences must be regular enough to define and separate entities. A feature is anything that can separate two entities in a consistent manner. Thus, differentiation can be seen as the detection of features unique to those entities.

2. **Impact of Feature Differences on Outcomes** – The differences in features must affect the outcome. Not all features will, but some must. For example, certain features in an organism's DNA affect reproductive success, while others do not. The challenge is to identify which features have an impact.

3. **Consistency of the Effect on Outcomes** – The impact of features on outcomes must be sufficiently consistent to allow for accurate predictions, reliable engineering, or contributions to survival.

When all three components are satisfied, we say a *Differentiation of Outcomes* exists, and a pattern is being followed. If we can precisely determine this pattern, clearly identifying the important features and their effects, we can exploit the pattern, make predictions, and create representations and simulations. If we desire a particular outcome, we can achieve it by manipulating the features of an event's interactions to produce the desired result.

For example, we can predict a cannonball's trajectory based on features such as the barrel's direction and the effects of gravity. We can control where the cannonball lands by adjusting the firing angle, gunpowder, and direction, provided we understand how these features influence the outcome. In chemistry, we can predict a reaction's result if initial conditions are known or produce specific chemicals by controlling variables like reactants, quantities, temperature, and pressure. The same principles apply across engineering.

II. A Theory of Patterns

The *Differentiation of Outcomes*, as described in terms of features and outcomes, is completely general. No specifics are given on how features are processed to produce an outcome. Which features are important? What weights do different features hold? How do features interact with each other? We essentially treat the calculation of features as a black box, with features being fed into the box and an outcome emerging.

This black-box approach allows us to generalize many phenomena. Processes such as science, machine learning, neural computation, and evolution can be seen as feature inputs into a black box, where calculations happen via language, mathematical formulas, computer algorithms, neural networks, DNA translation, or environmental selection in evolution. Outcomes can be predictions, categorizations, or reproductive success.

For a Patternist, *Differentiation of Outcomes* is the starting point for probing any phenomenon. What are the features, and what are the outcomes of an interaction? Are the features consistent, and to what degree? Are the outcomes consistent, and to what degree? Is there a *Differentiation of Outcomes*—does a change in features produce a change in outcomes? Is the interaction random? Are there hidden features or variables that have yet to be accounted for? Do we have the means to detect them, or must we develop new tools?

Is there a pattern being followed?

Once these questions are answered, we can examine how outcome calculations occur and how features are processed. We can "*open*" the black box to see the algorithms employed inside and extract out principles to replicate in another medium. By the end of this book, we will come to understand how all algorithms inside the black box—whether composed of biological or artificial neurons, silicon transistors, mathematics, natural selection, or even science—are essentially one and the same. They all operate on patterns, on the existence of regularities, and the *Differentiation of Outcomes*.

A Semi-Consistent Universe

In a perfectly consistent universe, the *Differentiation of Outcomes* would be followed exactly. Not only would every event have a cause and effect, but each event would possess measurable features that could be used to calculate its outcome with absolute accuracy. The result of every coin toss could be perfectly predicted, as the *Differentiation of Outcomes* would ensure that measurable aspects—such as the coin's launch angle, speed, and shape—would determine the result precisely. This would hold for every dice roll, every horse race, and every observable process. A perfect simulation of such a universe could run without any divergence.

This is what it means to live in a strictly consistent, deterministic universe. All patterns would be strict. The outcomes of all events would be perfectly predictable, and all events would have features that, once accounted for, would determine the outcome exactly.

We do not live in such a universe.

It is seemingly impossible. We are not omnipotent. We lack the capacity to measure, detect, and store the near-infinite number of features for a given event, including all previous states of the event that might influence the outcome. Nor do we have access to all external processes that might affect the event. Does the orbit of the moon affect the phenomena we are studying? It certainly affects the tides. What about the sun? The sun affects day and night, temperature, and the seasons. What about the solar system's revolution around the galactic core? Or a nearby supernova? How far do these external features go? To the outer edges of the known universe—or beyond? Even if we had access to all features, the computational power required to process them all would likely be beyond our reach.

We do not live in a perfectly consistent universe.

But this is not to say that we live in a completely random universe, either. There are regularities to identify and take advantage of. There are regularities that follow strict patterns, such as the fundamental laws of physics—unchanging and universal. There are regularities that exist momentarily and then vanish, like a song played on a piano. And there are regularities that exist on their own timescale, persisting before eventually fading. A desert becomes an ocean. An ocean becomes a desert. Seashells are found on the peaks of Mount Everest. An environment with all its regularities may last for millennia before it disappears due to gradual shifts or sudden floods, volcanic activity, or asteroid impacts. The same applies to living organisms: they exist as regularities in an environment for generations but can adapt, change, or suddenly become extinct.

We live in a semi-consistent universe.

There is randomness, and there is regularity. Some phenomena can be predicted with absolute certainty. Some phenomena can only be given probabilities. And some phenomena appear entirely random. This is the universe we currently inhabit. Debating whether the universe is ultimately deterministic or fundamentally random is pointless, as we have yet to reach that limit. We have not fully explored the universe's outer reaches, nor have we built an omnipotent supercomputer capable of detecting all features of all events, deciphering their relationships, and calculating their outcomes. For a Patternist—and for any living organism on Earth—the universe is both a random and a regular place.

Paradoxically, since randomness itself is an inherent regularity—a pattern within this universe—it too can be exploited by any intelligent system. Thus, randomness will be recognized as an essential principle in a *complete theory of intelligence*.

Medium Independence

Patterns are the reason the medium does not matter, enabling abstraction, representation, and language.

It is due to medium independence that mathematics and physics can occur. It is what gives power to symbols written using ink on paper. Mathematics relies on patterns of logic and structure, not on the physical medium through which those patterns are expressed. Equations and numbers are not bound to chalkboards or computers. The Pythagorean theorem describes a geometric relationship that a carpenter can use. Formulas for the volume of solids can be calculated on paper and translated into real-world objects. The Fourier transform provides the mathematical foundation for converting audio sounds into digital signals for storage and then reconstructing those signals to produce audio output.

In physics, the laws of motion, thermodynamics, and electromagnetism describe patterns of behavior that are universal. A falling apple and a collapsing star both follow the same gravitational principles, even though their mediums—Earth's surface and interstellar space as well as their material—are vastly different. The equations that describe these laws are medium-independent; they capture patterns that apply everywhere in the universe, regardless of the specific physical context.

Abstraction gains its strength from universality. By identifying the essential features of a pattern, abstraction removes the noise of the medium. This enables us to convey and manipulate ideas without being restricted by the form in which they originally arose, allowing the principles of a theory to be applied broadly. Representation, whether through maps, graphs, symbols, or neurons, utilizes medium independence.

Language operates on the same principle. Words—whether spoken, written, or signed—encode patterns of meaning. They make communication possible and allow ideas to travel across physical barriers, cultures, and generations.

II. A Theory of Patterns

It is through medium independence that biological neurons can be translated into artificial neurons.

It is through medium independence that there can be any hope for artificial intelligence. The belief that intelligence and consciousness are patterns—reproducible regardless of the medium—enables these phenomena to transcend their biological origins and be imbued into artificial systems.

As Patternists, our task is to identify the defining features that constitute intelligence and consciousness.

The Paradox of the Pattern

There is an inherent paradox within Patternism. We have defined patterns as discernible regularity, but this concept of *"discernible regularity"* is itself an oxymoron. It is paradoxical. How can anything be both *discernible* and *regular* at the same time? How can any *instance* of a pattern be both unique and repetitive?

All cats are the same, and all cats are different. All smartphones are the same, and all smartphones are different. All human beings are the same, and all human beings are different. Each of these statements is both arguably true and self-contradictory.

Is Patternism inconsistent, then?

The answer is yes.

But the trade-off for such inconsistency is completeness. Patternism is a theory that seeks to explain all biological phenomena, even if it means sacrificing consistency within the framework.

Although there are valid concerns about maintaining a strictly consistent theory, the paradox within the first principle of Patternism reflects the nature of reality and what can be built from it. This inconsistency, known as the *Paradox of the Pattern*, is found and utilized throughout nature. All living organisms make use of the *Paradox of the Pattern* to an immense degree, for it is only through the *Paradox of the Pattern* that the phenomena of reproduction, eating, and repairing can occur.

Consider a form of the *Paradox of the Pattern* in the ancient *Ship of Theseus* paradox:

A ship belonging to the Greek hero Theseus is kept in a harbor as a museum piece. Every year, a plank of wood from the ship rots and is replaced by a new one. After a thousand years, all the ship's planks have been replaced, raising the question: "Is this the same *Ship of Theseus*?"

II. A Theory of Patterns

The paradox continues: all the original planks that were removed from the ship are gathered and rebuilt in precisely the same arrangement. Now there are two ships, and the question is then asked: "Which of the two is the *Ship of Theseus*?" Is it the ship whose planks were gradually replaced, or is it the ship rebuilt with the original planks that were stripped away?

There are multiple proposed solutions to this paradox from different philosophical viewpoints. For a Patternist, the question of which ship is the *Ship of Theseus* centers on patterns—specifically, on representations and the criteria for naming something the "*Ship of Theseus.*"

For a Patternist, the *Ship of Theseus* is a pattern defined by a set of features. If an entity meets enough of these features to a certain threshold, the entity can be called the *Ship of Theseus*. However, this Patternist approach brings its own complexity. Who decides which features define the pattern of the *Ship of Theseus*, and what threshold must be met? Is it the universe by natural selection that decides? An authoritative figure like the museum director? A group consensus among visitors to the ship? A sailor who can operate it? Is it Theseus himself, making any ship he commandeers the *Ship of Theseus*? Or does the *Ship of Theseus*—the one Theseus sailed across the Aegean Sea in his adventures so long ago—no longer exists, much like a man is no longer the child he once was?

For a Patternist, the proposed solutions to the *Ship of Theseus* paradox are inherently inconsistent. It can be argued, from the fundamental principles of Patternism, that both ships are the *Ship of Theseus*, neither ship is, one ship is the *Ship of Theseus*, and the other is not.

But is this not true?

Is it not true that the solutions are inherently inconsistent?

The question of the *Ship of Theseus* can be asked about many different phenomena. It can be asked about *George Washington*. Is George Washington the skeleton in the tomb at Mount Vernon? Is he the painting hanging in the Oval Office? A wax figure in a museum? The letters he wrote, now stored in a library bearing his name? An actor in a movie? Or is he gone forever—and if so, why do we label all these things as *George Washington*? Who decides, and to what degree can an entity be called *George Washington*?

Even more astounding is that the *Ship of Theseus* paradox applies to DNA. When DNA replicates, half of the DNA strand is stripped from the other half, and a new complementary strand is built for each half using free-floating nucleotides. After one replication, we ask: "Which of the two set of DNA strands is the original?" Is it the leading strand, or the lagging strand? Is it both? Or neither? What about after millions of replications, as occurs over the lifespan of a multicellular organism? What about mutations? And if we consider replication across generations, with differences accumulating over billions of generations, what can we say about the DNA of a living descendant compared to its ancestor from a hundred million years ago? Which is the original DNA?

Again, through the *Paradox of the Pattern*: All nucleotide molecules are the same, and all nucleotide molecules are different. All DNA are the same, and all DNA are different. All living organisms are the same, and all living organisms are different.

The *Paradox of the Pattern* is universal.

Finding a precise answer to the Ship of Theseus paradox is not the point. For more than two thousand years since the paradox was first introduced, philosophers have been blind to only focus their attention on trying to find a definitive answer to whether the two ships are the same, which, even if an answer is provided, ultimately changes nothing. For a Patternist, the most critical aspect of the *Ship of Theseus* paradox is in the phenomena of reproduction, eating, and repairing that the paradox demonstrates.

In the *Ship of Theseus* paradox, a reproduction of the ship occurs. From one ship, a duplicate is made with all the functions and features of the original ship. But why stop there? Using the same method, we can take each ship, replace each plank, and use the old planks to build another set of ships, resulting in four ships of Theseus. And why stop there? We can take each of the four ships, replace each plank, and rebuild to create eight ships of Theseus. Then sixteen ships. Thirty-two.

Sixty-four. One hundred and twenty-eight. An entire fleet of ships of Theseus can be built using this method.

What the *Ship of Theseus* truly illustrates is the beginning of **reproduction**. For more than two thousand years, philosophers have overlooked this critical aspect of the *Ship of Theseus* paradox, even though the same process is repeatedly observed in nature. It is what occurs when an egg is fertilized by sperm, and the resulting zygote divides from a single cell to many. It is what occurs in mitosis, within all multicellular organisms as they grow and age. It is what occurs in manufacturing, on assembly lines, and through the standardization of component parts, resulting in the mass production of cars, airplanes, and computers.

The second aspect that the *Ship of Theseus* paradox demonstrates is the phenomenon of *eating*. In the paradox, an old ship is technically broken down into its individual components—its planks and nails—which are then used to build a new ship. This process can be generalized further. Since planks and nails are component parts of many objects, the new ship does not have to be built from the old ship's parts. A *Ship of Theseus* can be built from planks and nails taken from several smaller ships, from a wagon, from a house, from anything made up of wooden planks.

Again, the crucial aspect of the paradox is that this mirrors what occurs in nature. When an organism eats another organism—when a predator consumes prey—the predator breaks down the prey into its component parts, into a mush of nutrients that the predator uses to draw energy and rebuild itself.

All organisms eat. All organisms engage in the process of gathering nutrients from the environment to rebuild and eventually reproduce themselves. Philosophers eat. Scientists eat. Musicians, kings, presidents, and cognitive scientists eat. All human beings eat. We eat plants. We eat animals. And, in desperate times, we eat each other. The phenomenon of eating is seen everywhere in nature, and yet it has been largely ignored.

At the core of the *Ship of Theseus* paradox is a practical question about the phenomenon of **repair**. In the paradox, the original ship of

Theseus, stored in the museum harbor, decays and becomes damaged over time, requiring annual repair and maintenance. This raises the question: "What does it mean for an entity to be repaired?" For a Patternist, an entity is repaired when its original features are restored to their initial values. In the paradox, an old plank of wood in the ship no longer retains its original features—such as hardness or elasticity—so the old plank is replaced by a newly cut and cured plank.

Again, there are nuances regarding who gets to decide what those features are. A sailor might prioritize the ship's ability to sail without sinking, while the museum director might consider the ship repaired only when it matches the descriptions left by the previous director. Even the ship itself, if it were conscious and had a representation of itself to compare to, can decide when repairs have been made.

The act of repairing is not limited to the *Ship of Theseus*; it occurs throughout nature and can be seen in phenomena like eating, as well as in all manufactured products. When a car breaks down, we take it to a mechanic for repairs. The mechanic might replace the engine or tires, restoring the car's key feature: its ability to drive. The same principle applies to computers, tools, and other machines. The question posed in the *Ship of Theseus* paradox extends to any repaired object: "Is a repaired car the same as it was before?" Is a laptop still the same after most of its parts have been replaced and upgraded? Is a remodeled house still the same house? Is a man the same person he was as a child, even though every cell in his body has been replaced many times over as he ages?

The phenomena of reproduction, eating, and repairing all depend on patterns and the *Paradox of the Pattern*. The *Ship of Theseus* paradox requires that planks of wood be consistent regularities within the environment. All planks must be similar enough to be interchangeable, yet different enough that replacing an old plank with a new one has meaning. For eating to occur, the predator and prey must share compatible biochemistry, allowing the predator to break down and absorb nutrients from its prey. In organ transplantation, organs like hearts, kidneys, lungs, and livers must also exhibit regularity across the population. One heart can replace another, provided the essential features match and the heart is not rejected by the recipient.

Again. All ships of Theseus are the same, and all ships of Theseus are different. All cars are the same, and all cars are different. All kidneys are the same, and all kidneys are different. The *Paradox of the Pattern* is inescapable.

Why have philosophers failed to notice the phenomena of reproduction, eating, and repairing embedded within the *Ship of Theseus* paradox? Why has no one recognized that these phenomena are deeply connected, share an underlying principle of patterns, and can be extended to explain phenomena beyond the biological substrate, serving as clues to a theory of intelligence? As of the writing of this book, why is there no theory that seeks to generalize and account for these essential and immensely practical phenomena— phenomena that every human being and every organism has encountered and engaged in daily since life first emerged over four billion years ago?

It is inexcusable and a massive failure by philosophers and scientists to be blind to such fundamental phenomena. It is a failure driven by the human-centric view of intelligence, which causes philosophers and scientists to focus exclusively on human-associated phenomena—abstraction, representation, language, naming, and identity—while neglecting the essential processes that all organisms participate in.

The human-centric view of intelligence is not only wrong but is an impediment to finding a *complete theory of intelligence.*

A Patternist will make no such mistake. We will start from the very bottom and climb our way up. Reproduction, eating, and repairing form the foundational rungs of the *Ladder of Complexity* to which consciousness will be reached.

Any theory that seeks to explain intelligence, evolution, or technological innovation must first account for the fundamental phenomena of reproduction, eating, and repairing. Any theory that fails to do so, that cannot address these essential and highly practical phenomena, is ultimately a waste of time.

We have shown that at the core of the phenomena of reproduction, eating, and repairing lies the *Paradox of the Pattern*. But the *Paradox of the Pattern* covers so much more. All paradoxes and inconsistencies found in nature can be traced back to patterns and the *Paradox of the Pattern*: the paradox of self-referencing, the problem of how a species is defined, the question of how a representation can be both the same as and distinct from the entity it represents. The question of why children resemble their parents but are also different. The shifting boundaries of language and definitions and their manipulations for propaganda. Perception and illusions. The spontaneous generation of cancer cells in multicellular organisms. History, as both the recording of new events and a repetition of the past. Debates over morality. The human tendency to form tribes, wage war, commit acts of ethnic cleansing and genocide. Even in politics, between those who strive to uphold traditions that have sustained society and those who aim to reform and improve the system. All can trace their roots back to the *Paradox of the Pattern*.

All is one. One is all. One is one. One is nothing. And all is nothing. Each of these statements can be argued as true when taken to the logical extremes within Patternism. Consequently, Patternism is inconsistent. But this inconsistency within Patternism reflects the inconsistencies found in nature. Is it not true that we can trace our ancestry to a single-celled organism? That our atoms originated in the cores of stars? Is it not true that everything and everyone—even the sun itself—will eventually change and disappear? Is it not true that each of us is unique? That we can only perceive reality through our own eyes? And is it not true that, at times, what we consider our self—our actions and thoughts—seems beyond our control, almost an illusion? That the reality we experience in dreams is not real?

For things to be different and yet the same.

For things to be the same and yet different.

All cats are the same, and all cats are different.

All human beings are the same, and all human beings are different.

All patterns are the same, and all patterns are different.

II. A Theory of Patterns

A solution to the *Paradox of the Pattern* will not be answered or offered here. For a Patternist, patterns and the *Paradox of the Pattern* are inherent properties of the universe. As it stands, the *Paradox of the Pattern* will remain the fundamental mystery within the heart of Patternism.

III. The Search for Patterns

Adaptive Systems

Reproduction, eating, and repairing cannot occur without patterns. Language, representations, and mathematics rely on patterns to operate. Patterns are the reason why symbols written using ink on paper can reflect reality. Predators search the environment for patterns leading to prey, while prey animals are on the lookout for any loud noise or irregular scents—patterns that indicate predators are nearby. Patterns serve as principles in scientific theories.

Given their vast range of uses, patterns are undeniably valuable.

But how are patterns identified and stored? What are the underlying principles, and what mechanisms can be engineered to find and store those patterns?

Due to their immense usefulness, systems capable of automatically identifying and storing patterns have evolved naturally. Any system that can learn and adapt is a system that has found a pattern in the environment and taken advantage of it.

The adaptive human immune system detects regularities in foreign pathogens and creates antibodies specific to those patterns. Human memory stores past events, extracting patterns to predict future events. Genes store patterns by preserving traits that have exploited patterns in the environment for survival and reproduction. The muscular system of an athlete stores effective movement patterns through training.

In addition to naturally evolved pattern-finding systems, there are man-made systems. Language, oral history, and writing capture patterns that have emerged over generations, passing knowledge across individuals, time, and space. Science, along with institutions like universities and libraries, is designed to rigorously identify, verify, and store patterns found in nature. Artificial intelligence and machine learning employ computers to uncover patterns in vast datasets.

In general, any *Adaptive System* capable of finding and storing patterns will have three distinct components:

1. A *randomizer* that generates variations to explore the environment.

2. A *sorter* that detects when a pattern has been found, evaluates its usefulness, and decides if it should be stored.

3. A *flexible medium* to store representations of discovered patterns.

Adaptation is intelligence. The human-centric view equating intelligence solely with consciousness must be discarded. Instead, any system that can learn—can find patterns in its environment and use those patterns is intelligent.

Viruses qualify as adaptive systems. A virus's flexible medium is its genetic material. It generates variation through replication errors and mutations. It sorts through its protein shell, which determines if a host has the necessary replication machinery, ultimately leading to natural selection. By this definition, viruses are intelligent.

Limiting intelligence to human beings is not only unproductive but dangerous. The history of viral diseases reveals the cost of such a narrow-minded view. Consider the hundreds of millions who have died from smallpox, Ebola, and influenza. How do these viruses mutate, evolve, and develop drug resistance over time? Why do governments spend billions monitoring outbreaks and responding when they inevitably occur? And this is only viruses. If we include bacteria and protozoa—the agents of diseases like plague, typhoid, and malaria—the death toll easily climbs into the billions, accompanied by staggering treatment costs.

And what about the weaponization of pathogens? Are we to ignore the reality of biological warfare? Why do we engineer viruses to be more lethal and contagious, and invest heavily in countermeasures? Why do we use them to kill other human beings?

The human-centric view of intelligence is fundamentally wrong.

If we claim that disease causing viruses, bacteria, and protozoa are not intelligent, then what good is a definition of intelligence? What good is a definition that is impractical and doesn't help us survive? What good is a definition that applies only to human beings and fails to account for the immense impact of microorganisms on our world?

For a Patternist, the ability to find and exploit patterns is intelligence. Intelligence is not exclusive to consciousness or thought but rather the capacity to adapt, survive, and exploit opportunities, and it exists in countless forms across nature. Ignoring this is underestimating both our vulnerabilities and the intelligence in systems beyond our own.

Do not underestimate adaptive systems. The same adaptive systems that pose threats can serve as powerful tools. Human beings have evolved our own adaptive systems to counter viruses and bacteria.

The adaptive immune system and its ability to learn is what makes vaccinations effective. Vaccines work by training the immune system to recognize specific patterns of a virus through exposure to a weakened or inactive version. This allows the immune system to practice identifying the virus without risk. When it later encounters the live virus, it can immediately recognize it as a threat, releasing the correct antibodies to neutralize the virus before it can cause harm.

But how exactly do adaptive systems work? How does a virus adapt to a host, and how does the immune system adapt to pathogens? What general principles are involved? Seeking answers to these questions provides a more universal theory of intelligence than the human-centric approach focused solely on human cognition.

While we have reduced adaptive systems to three core components—a randomizer, sorter, and flexible medium—pinpointing where these components operate and where adaptation occurs can be challenging. In the case of viruses, the adaptive system is the viral population as a whole, since individual viruses lack the ability to adapt on their own. More precisely, it is specific lineages within the population that evolve, as viruses cannot communicate with one another. Returning to the *Paradox of the Pattern*, we can say that all viruses are the same, and all viruses are different; no single virus adapts, yet collectively, the viral population adapts.

The same complexity applies to genetics and biological evolution in general. It is the population of organisms that adapts, with the flexible medium involving not just the genetic material of a single individual but multiple redundancies across the entire population. Moreover, adaptive systems can operate on multiple levels of randomization. In evolution, random mutations occur at the genetic level, with additional layers of randomization introduced through meiosis, sexual reproduction, and the random encounters between male and female organisms in the environment.

Adaptive systems are not isolated; they can stack and cooperate. When we examine human beings as a whole, we find multiple levels of adaptive systems operating—from the microscopic level to the individual level, to the population level, and beyond. These adaptive systems include our genes, immune system, muscle memory developed through training, the human brain, language, history, science, and more. Each system works within its own timeframe and is optimized to find and store patterns within its respective domain.

Just as adaptive systems can work cooperatively, an individual adaptive system can become corrupt, operating independently and propagating itself at the expense of the surrounding systems it once worked with. An example of this behavior is seen in the phenomenon of cancer cells.

A multicellular organism is a conglomeration of trillions of adaptive systems working together for the benefit of a single overall organism. Every cell in a multicellular organism is technically an adaptive system, containing a flexible medium (genes), a randomizer (mutation during mitosis), and a sorter (natural selection). Thus, every lineage of a cell has the capacity to learn and adapt, as each has all the necessary tools to do so.

III. The Search for Patterns

Under normal conditions, the adaptive mechanisms of cells are suppressed. Systems have evolved to ensure each cell's integrity and prevent corruption during cell division. However, under certain conditions—such as exposure to carcinogens or radiation that increases randomization—cellular adaptive systems can be triggered. Combined with evolutionary selective pressures, in which each cell reproduces and propagates itself, cells that were once benign can become corrupt, adapt, and turn cancerous, spreading as cancer cells that exploit their benign surroundings.

Cancer is spontaneous. Unlike infectious diseases, cancer originates when an organism's own cells divide uncontrollably, compromising the health of the overall organism. These cells form a population capable of surviving outside the organism's main body if transferred to a suitable environment. In this way, cancer cells can be seen as distinct living organisms—a new, self-propagating pattern.

The ability to adapt can lead to both cooperation and corruption. The emergence of multicellularity exemplifies adaptive cooperation, while the rise of cancer cells illustrates adaptive corruption. Similarly, the immune system, designed to identify pathogens, can become corrupted, reacting to harmless substances and resulting in allergies, severe overreactions, or even attacking healthy cells, leading to autoimmune diseases.

In Patternism, cooperation and corruption within adaptive systems are generalized. Phenomena such as symbiosis—whether mutualistic, commensalistic, or parasitic—as well as cancer, predation, genocide, ethics, morality, political and organizational corruption, and even religious and ideological conflicts, are all manifestations of adaptive systems interacting and adapting to one another.

For a Patternist, it is insufficient for a theory to explain a single phenomenon; it must seek to explain all related phenomena. In our exploration of adaptive systems, we have broken down systems capable of learning into the components of a randomizer, sorter, and flexible medium—not merely to be reductionist, but to enable prediction and engineering.

Recognizing that adaptive systems contain these components provides a powerful framework for analyzing any system that exhibits signs of intelligence and learning. We can identify the flexible medium that stores information and knowledge, locate the randomization component that generates variation, and pinpoint the sorter that determines when a pattern is recognized. If our goal is to design new adaptive systems, we can start by developing these individual components, combine them, and then proceed accordingly.

III. The Search for Patterns

Pattern Recognition

In our discussion of adaptive systems, we assigned all pattern recognition tasks to the sorter component. But how does a sorter recognize when it has encountered a new pattern? More broadly, how is a pattern recognized?

We observe that the adaptive immune system detects patterns in pathogens. The human brain rapidly identifies patterns both spatially and temporally. Through evolution, populations of organisms recognize environmental patterns and store them in the genetic pool. In science, naturally occurring patterns are rigorously documented. By examining these adaptive systems—as well as systems such as neural networks, evolutionary algorithms, and muscle training—we identify four general mechanisms that automatically recognize patterns: *Repetition, Prediction, Natural Selection,* and *Categorization*.

Mechanisms Underlying Pattern Recognition:

1. **Repetition** – Repeated occurrence indicates the existence of a pattern.

2. **Prediction** – The ability to make accurate predictions indicates that an entity has identified a functioning pattern.

3. **Natural Selection** – Consistent survival and propagation demonstrate successful utilization of a pattern.

4. **Categorization** – *The Differentiation of Outcomes* and the consistent grouping and separation of entities into sets and subsets, leading to naming and labeling.

We define a pattern as a discernible regularity composed of a set of features, where a feature is anything that can consistently distinguish between two entities. In pattern recognition, the goal is to identify the specific set of features that define a given pattern.

At a higher level, we examine the concept of the **meta-pattern**, which describes the features of patterns in the most general sense. The mechanisms of pattern recognition can be thought of as the *"features of a pattern's features,"* serving as criteria for filtering important features from unimportant ones. Essentially, a pattern is composed of features that are repeated, features that enable accurate predictions, features that aid in survival and reproduction, and features that facilitate effective categorization.

The set of mechanisms described here may still be incomplete. Processes such as abstraction, compression, and algorithmic generation can also produce patterns. Our goal is to identify the main mechanisms that enable the automatic recognition of patterns and to generalize these mechanisms for use within any medium. If a new adaptive mechanism is discovered, it should be added to the list.

For a Patternist, there may be additional mechanisms beyond *Repetition*, *Prediction*, *Natural Selection*, and *Categorization* that can aid in pattern recognition.

Mechanism 1. Repetition

Patterns can be detected through repetition. Since a pattern is a discernible regularity, the features that make up a pattern should naturally be regular and detectable through their repetition. The key is to build mechanisms that can automatically detect repetition, triggering a signal within the adaptive system to store the features that were present when the repetition occurred.

Human muscle training, or athletic training, is an adaptive system that utilizes the mechanism of repetition to store effective patterns of movement. Reinforce what is constantly in use. Athletic training—whether in martial arts, team sports, individual sports, or simply weightlifting—is adaptive in that the human body learns through repetition to coordinate muscles in ways that produce specific movements, from hitting a ball with a bat to playing the piano.

At a micro level, this adaptation occurs through the reinforcement of muscle fibers and neural pathways that are repeatedly activated, creating a positive feedback loop: the more a movement is practiced, the more efficient it becomes. Combining this reinforcement with a goal-recognition system that identifies successful movements further accelerates learning. Training can also be enhanced through the use of rewards and punishments, tapping into an agent's natural goal-seeking and avoidance behaviors. This principle is demonstrated in classical conditioning, as seen in Pavlov's dog experiments or the training of police animals, where stimulus and repetition—such as a command word or the ringing of a bell—are used to instill a desired response, training the animal to follow commands.

This process of learning through repetition begins with the athlete having limited knowledge, coordination, or timing for a particular movement. Guided by a teacher or coach, through visual feedback from observing others, or simply through trial and error, the athlete varies their movements until a successful action is achieved. The athlete then strives to replicate the movements that led to the initial success, focusing on reproducing all relevant features present in that moment. With continued practice, the athlete's body adapts, isolating the essential features of the movement and eventually achieving efficiency without requiring additional cognitive effort.

The human brain alone cannot command the body to perform unfamiliar actions with competence without practice. We cannot simply will our bodies to perform novel actions perfectly; the body must engage in practice. Although this is a simplified view of the human musculoskeletal and neuromuscular systems and their roles in adaptation, the principle of repetition in practice is undeniable. Without the automatic reinforcement of frequently used cells, there can be no adaptation and no learning of new movements within an individual's lifetime.

Athletic training—or, more broadly, the art of practice—is an intelligent process. If we consider the art of practice to include the teachers, students, coaches, various styles and schools, along with the introduction of novel techniques found through individual experience and the spreading of those techniques through books and seminars, from teacher to student, and the testing of effectiveness through sparring, competition, or commercial success, we see that the art of practice contains all the components of adaptive systems and mirrors the process of evolution.

Practice must be viewed in terms of patterns. There are regularities within each specific activity that can be taught to an individual starting from scratch. It is through this process that a baby learns to walk, a child learns to ride a bike, and we learn to swim, drive, or write. It is through the art of practice that sports teams win championships, gold medals are earned, and experts emerge. The practical value of the art of practice is immense. At its core, what enables the art of practice to occur and allows patterns of movement to be perfected is the automatic detection of features by the mechanism of repetition.

For a Patternist, the art of practice is a biological phenomenon and a key to understanding intelligence, just as the adaptive immune system and biological evolution are. Why do people go to gyms? Why do we enroll our children in soccer, violin, or piano lessons? Why do

III. The Search for Patterns

we do homework or solve problems from a textbook? What is the purpose of lifting weights? What does it mean to train muscle memory?

For a Patternist, studying the art of practice reveals the mechanism of repetition that can be applied to other fields. It can be applied in engineering to build robots that automatically learn how to move. It can be applied to the human brain to understand how networks of neurons find patterns, showing that the mechanism of repetition underpins how neurons are wired to identify and reinforce the regular features of a pattern.

Learning through repetition is not without its disadvantages.

There is a substantial divide between a pattern and the recognition of a pattern. What we can store is only the recognition of a pattern, not the pattern itself. In the process of storing the recognition of a pattern through repetition, mistakes can be made. Errors during the learning process of practice can form bad habits that impede one's abilities. Once ingrained, old habits are difficult to unlearn and may interfere with new learning.

In the human brain, the wiring of neural networks through repetition of features can lead to bias if the training sample is too small or if relevant features are missed. Patterns can be misinterpreted or falsely stored, creating recognition errors that can lead to confirmation bias and hallucinations.

Repetition can also give rise to deception. The human brain is wired to notice repetitions and learn from them. This becomes a pattern to exploit. Keywords and phrases are repeated into slogans by the media and government, giving rise to the phenomenon of propaganda. These repetitive messages become deeply embedded and difficult to dislodge within the target population. Through propaganda, repetition becomes a tool for control, as human beings act not on what is objectively true but on what we perceive to be true.

We act on the patterns that we have learned.

Mechanism 2. Prediction

Patterns can be identified by making accurate predictions.

All predictions—whether forecasting the weather, anticipating an opponent's next move in chess, or calculating the path of an object in physics—are generated from patterns. If the patterns we use are accurate, meaning the correct features have been identified and properly utilized, then the predictions will also be accurate. Conversely, if we consistently make accurate predictions, we can be confident that we have discovered and utilized an underlying pattern, even if the initial discovery of the pattern was entirely random.

In essence, prediction accuracy can serve as a selection factor in an evolutionary algorithm to identify patterns automatically. A potential pattern generates a prediction, which is then compared to the actual outcome to assess its accuracy. This accuracy is used to determine whether the potential pattern qualifies as an actual pattern.

Since pattern recognition is fundamentally a feature-filtering process, any random prediction must still incorporate features of an initial event. What we seek to avoid is mere guessing, where no features are considered at all.

The Prediction Process involves five steps:

1. *Capture the Initial Features* of an event.

2. *Feed the Initial Features into a Prediction Engine*.

3. *Generate a set of Prediction Features*—a detailed description of what is expected—from the *Prediction Engine*.

4. *Observe the Outcome* and collect the *Outcome Features*, which describe what actually occurs.

5. ***Compare the Prediction Features to the Outcome Features*** by matching features. The more features that match, the more accurate the prediction.

These steps are designed to be general and universal, applicable to all phenomena where prediction occurs. For instance, a scientist makes predictions using a theory and conducts experiments to test the theory's accuracy. Similarly, a poker player predicts an opponent's hand based on visible cards, bets placed, and facial expressions. Whether predicting the existence of black holes, the path of Venus, or the next word in a sentence, all predictions follow these steps in some form.

When can predictions be made?

Before making a prediction, we must first ask: "Is it even possible to make a prediction about an event?" For a Patternist, predictions can only be made if a pattern is being followed and there is a *Differentiation of Outcomes.* If an event is governed by an underlying regularity, its outcome will also be consistent and therefore predictable. Conversely, if an event is entirely random, no accurate prediction can be made.

If we assume that events and their outcomes are not entirely random, we assume that there is an underlying pattern to be found. This pattern should contain a set of features that we need to isolate and identify. This is the difference between legitimate predictions and blind, random guesses. Legitimate predictions are reproducible and can be taught to others. They operate on the belief that certain features of the current event will influence the outcome, and these features must be incorporated into a calculation to generate a prediction. Blind, random guesses ignore features entirely, generating predictions without any inputs. This brings us to the first step of the prediction process: capturing the *Initial Features* of an event.

Initial Features are simply information about the current event that we feed into our *Prediction Engine*. These features can include quantitative, qualitative, visual, audio, or temporal data—descriptions, recordings, photographs, or any information that can consistently distinguish the event from others.

The main challenge is ensuring that the critical features of the governing pattern are included in the data we collect. This requires appropriate tools, sensors, and equipment. Casting a sufficiently wide net increases the likelihood of capturing essential features, enabling irrelevant features to be filtered out later. Failure to capture these critical features compromises our ability to make accurate predictions, leaving us effectively blind and rendering outcomes unpredictable.

Once we have a set of *Initial Features*, we feed them into a *Prediction Engine* to generate a prediction. A **Prediction Engine** is any system that takes features as input and produces a corresponding set of features as output. By this definition, mathematical equations, scientific theories, field guides, and even the human brain qualify as prediction engines.

The *Prediction Engine* can be treated as a black box. We do not need to fully understand how the input features are processed or combined to produce the output prediction. As long as the predictions are consistently accurate, we can be confident that the engine has identified and utilized an underlying pattern. However, if we understand the internal mechanisms of the *Prediction Engine* and can clearly describe the principles acting on the input features, then accurate predictions provide validation of the principles used to generate the prediction.

Prediction Engines can also be used to create simulations by taking their output features and feeding them back as input features, creating a loop. This looping process forms the basis of what we recognize as simulators. A physics engine in a computer program, a video game generating a fantasy world, or the way a chess player mentally simulates a chessboard to explore possible moves—all use this looping process.

Using a prediction engine as a simulator comes with trade-offs. As the simulation loops and runs, the accuracy of predictions tends to degrade. The further forward in time or the more steps the loop takes, the more likely errors will accumulate—even if the initial prediction was highly accurate. Additionally, simulations built on inconsistent

Prediction Engines may exhibit chaotic behavior as they run, even when starting with the same initial input features.

How accurate does a *Prediction Engine* need to be?

A *Prediction Engine* can be highly complex and variable. It may be strict, generating only a single prediction per input step, or it may produce a range of predictions or scenarios for each input. As long as the prediction engine captures the critical features of the underlying pattern, integrates those features effectively, and avoids blind guessing, it has the potential to generate accurate predictions.

How is the accuracy of a prediction determined?

How can we tell if a map we are following has led us to the correct location? How does a store manager ensure that the inventory list stored in a computer program matches what is physically available for sale? How does a physicist determine the velocity of an electron in a given magnetic field? How does a person climbing a flight of stairs know where to step, and what happens when a step is missed?

In all cases, prediction accuracy is determined by comparing *Outcome Features* to *Prediction Features* and counting the number of matches. The more features that match, the more accurate the prediction. Conversely, fewer matches indicate lower accuracy. Additionally, by running the *Prediction Engine* multiple times to generate and check multiple predictions, we can assess whether the engine is consistently accurate overall.

This process of checking prediction accuracy can be implemented mechanically. Neurons in the human brain operating on threshold potentials, neural networks, if-else statements, logic gates, and even language used to describe outcomes are all tools that can be designed to compare features and automatically compute prediction accuracy.

The specific features compared depend on the type of sensors available to the prediction system. If the system has visual sensors, the *Prediction Engine* can generate a visual prediction to compare with outcome features detected by those same visual sensors. Similarly, audio sensors enable audio-based predictions, and so on. This approach applies across sensor types, whether they detect infrared, temperature, atmospheric pressure, or radiation.

However, the focus should not be solely on the accuracy of a single prediction but rather on comparing accuracy across multiple predictions—either from different prediction engines or from the same *Prediction Engine* after introducing variations. Such comparisons allow us to determine which *Prediction Engine* performs best in terms of prediction accuracy. In essence, this provides a gauge for testing the fitness of various prediction engines. This methodology enables the implementation of an evolutionary algorithm to optimize for the best *Prediction Engine*—the one that most effectively captures the underlying pattern.

Using prediction accuracy as a selection factor to identify patterns has been a foundational approach in science. All scientific theories function as prediction engines: when a theory produces accurate predictions, it confirms that the principles underlying the theory reflect actual patterns in nature. Science, as a whole, is an evolutionary process that selects prediction engines based on the accuracy of their predictions. Scientific progress is the result of evolutionary pressure, with the theories that survive being the best predictors, while the rest are discarded and forgotten. It is precisely because of this automatic mechanism of selecting theories that science is such a powerful and objective tool.

The belief that progress in science is driven solely by human reasoning is a lie. Our role is simply to create and vary prediction engines. For a Patternist, the opposite is true: science progresses in spite of human interference and human nature. Individual ego, corruption, politics, and the tendency to rationalize are all present within academic institutions, as they are elsewhere in any organization involving human beings. But because prediction accuracy can be mechanically calculated and made undeniable, obstacles to progress arising from the flaws of human nature can eventually be overcome. Current theories evolve and compete with new theories. If they cannot, they die, and the new theory takes their place despite any protest—just as it occurs in nature and everywhere else.

III. The Search for Patterns

At the core of the evolutionary algorithm is variation, which leads to outcome differentiation. Variation in scientific theories arises from the unique experiences of individuals, with each theory developing distinct principles that can then compete for prediction accuracy.

But where, in general, can variation occur in a prediction engine? In the prediction system we have described, there are two main points where variation can be implemented.

Points of Variation in Prediction Engines:

1. **Input Features** – Varying the range of input features increases the likelihood of capturing critical features.

2. **Internal Processing** – Variation can also occur in how input features are processed within the prediction engine.

The first point of variation involves the type of features fed into the *Prediction Engine* as initial inputs. Since the number of features that can be input is limited—yet there are near-infinite possible features for any given event—the system's task is to capture the critical features of the underlying pattern within the system's restricted bandwidth. Varying input features increases the likelihood of incorporating important elements. Once these critical features are correctly identified, the engine can be optimized to focus exclusively on them, filtering out irrelevant features. This search for an effective input spectrum might involve recalibrating existing sensors to explore different areas or developing more advanced detection tools—such as telescopes or microscopes—to capture new features, improve prediction accuracy, and identify increasingly detailed patterns.

The second point of variation lies within the *Prediction Engine* itself, specifically in how input features are processed to generate predictions. When we vary the internals of the engine, we rearrange and manipulate its components to produce a calculation, ultimately generating a prediction. In science, this variation can be methodical, involving mathematical reasoning, data organization, testing new calculations, or seeking regularities within observations. In other prediction engines, such as neural networks, variation can be more random, with connections between neurons randomizing to create new pathways and thus altering how predictions are calculated.

In an evolutionary algorithm, variation is iteratively applied to the *Prediction Engine* that currently yields the best results, building on prior successes. Similar to biological evolution, where distinct species can be traced, lineages of prediction engines can become specialized over time. As in evolution, a balance in varying a *Prediction Engine* is crucial to ensure the search proceeds effectively. If variation is too great, the features that contribute to prediction accuracy may be lost. Conversely, if variation is too limited, the prediction accuracy may become trapped at a local maximum, unable to break free to discover features that would lead to greater accuracy.

A major philosophical question arises when we examine how a *Prediction Engine* processes its input features. So far, we have treated the internals of a *Prediction Engine* as a black box—taking in features as inputs and outputting a set of features as predictions. What happens between input and output is essentially a function. The key question to ask is: "Are all functions computable?"

This is a complex and challenging question. A simpler question is: "Can all phenomena in the universe be represented?" Similarly, are all phenomena in the universe describable by science? Can all environmental patterns be captured in DNA?

If the answer is *"yes,"* it implies that we can create representations for all possible phenomena and, by manipulating these representations, predict all possible outcomes. If the answer is *"no,"* we are left to ask: "What are the limits of representation?" What subset of phenomena in the universe can be predicted?

In the same way, when it comes to human intelligence, we must ask: "Can all phenomena be represented by networks of neurons?" If yes, then by manipulating those neurons—even somewhat randomly—accurate predictions can be made. If no, certain phenomena will remain unpredictable and beyond human comprehension, as all knowledge and understanding must be filtered through the human brain and its neural networks.

The full implications of this philosophical question will be explored later. For now, we adopt the Patternist perspective: *"If it follows a pattern, then it can be represented. If it follows a pattern, then it can be predicted."*

If something is regular, it can be named. If it is regular, it can be represented as neurons in the brain. If it is highly regular, it can be calculated using mathematical symbols. By varying those symbols, modifying language, or altering the connections of neurons, we can employ these systems to make predictions.

This philosophical inquiry into the computability and representation of phenomena directly ties into the practical role of prediction in identifying irregularities. If we accept that patterns can be represented and predicted, it follows that deviations from these predictions—***irregularities***—signal underlying phenomena not yet accounted for by the current system.

Prediction then becomes more than a tool for anticipating outcomes; it becomes a mechanism for detecting the existence of irregularities.

For an intelligent agent, recognizing irregularities is critical to its ability to adapt and survive. An irregularity indicates that the agent has encountered something new—an anomaly that current beliefs, theories, or systems fail to explain or defend against. It signals that the patterns the agent relies on are incomplete or outdated, necessitating relearning or the discovery of new patterns.

But how does an agent recognize an irregularity? Using the prediction process described, we can employ a well-trained *Prediction Engine* as an ***Irregularity Detector***—a mechanism that automatically triggers when an irregularity is encountered.

This process begins with programming a well-trained *Prediction Engine* to continuously compare its predictions against the current environment. Since a well-trained *Prediction Engine* has already captured the underlying patterns, its predictions are typically accurate under normal circumstances. When the *Prediction Engine's* output aligns with reality, the agent can relax and conserve energy. However, if the engine's prediction suddenly fails, this signals the presence of an irregularity, prompting the agent to initiate specific responses.

The response can vary. The agent might increase its focus, shift out of low-energy modes, or expend more energy through its sensors to investigate the irregularity's source. If predators are known to exist in the area, the agent might go on high alert, ready to respond. If the agent is truly intelligent, encountering an irregularity should activate its learning modules, preparing it to learn a new set of patterns.

For instance, as we walk, we unconsciously predict where our next step will land, expecting a stable floor beneath us. This allows us to focus on other tasks while our bodies move and adjust. If we suddenly take a step and do not feel a floor where we predicted it to be, we snap back to alertness, trying to understand what went wrong. Was the floor uneven? Was there a change in height? A hole? Did we trip over an object?

This phenomenon, where the detection of irregularities prompts an adaptive system to reevaluate its current beliefs, is evident throughout science. Wave-particle duality emerged from irregularities observed in the double-slit experiment, while Einstein's special relativity arose from the failure of aether theories to explain the Michelson–Morley experiment's measurements of the speed of light.

Irregularities that occur consistently are the strongest indication that a pattern exists to be discovered. Such consistent irregularities must be fully investigated by further probing to reveal their features, until the irregularity itself becomes a recognizable, usable pattern. This new pattern is then incorporated into the intelligent system, updating both the prediction engine and the irregularity detector in an endless cycle—a cycle that continuously validates and refines existing patterns.

III. The Search for Patterns

Mechanism 3. Natural Selection

Natural selection is a universal mechanism that identifies patterns by filtering them from the environment. The patterns it selects are solutions that directly or indirectly enable an entity to survive and reproduce.

The current theory of evolution is incomplete.

The belief that the current theory of evolution—centered on reproductive success and population genetics—fully explains all biological phenomena is false. Viewing natural selection solely in terms of variation, inheritance, differential survival, and reproduction fails to address a deeper question: "What exactly is it that survives from one generation to the next?"

Is it the individual organism? But the individual organism dies.

Is it the offspring? But the offspring will also die.

What, then, truly survives? Furthermore, what exactly is reproduction? How does it occur, and is it solely about quantity? For instance, in social organisms like ants, only the queen reproduces while the workers remain sterile. In multicellular organisms, every cell reproduces, yet only the germ line—the fertilized egg—continues to the next generation, while trillions of other cells eventually perish.

The current theory of evolution cannot answer these fundamental questions because survival and reproduction are taken as first principles—the foundational assumptions on which the theory is built.

All theories operate by accepting a set of principles as true, from which everything else is derived. It is then reasonable for the current theory of evolution to rest on the axioms of survival and reproduction. These principles allow for the explanation of many biological phenomena and enable numerous predictions. However, to claim that all biological phenomena can be explained solely by variation, inheritance, differential survival, and reproduction—that the field of biology is complete—is false.

Patternism seeks to replace the current theory of evolution not because the current theory is wrong, but because a deeper level of understanding can be achieved. Patternism views evolution, living organisms, and natural selection in terms of patterns. Survival and reproduction are secondary phenomena that emerge from the recognition, utilization, and propagation of patterns.

For a Patternist, *Natural Selection* is a pattern-filtering mechanism that enables organisms to automatically identify patterns in the environment and store their recognition as genetic representations.

What living organisms discover through the process of natural selection is not just physical traits that help in reproductive success, but solutions to maintain existence. What is found are means to which an entity can maintain the pattern of itself and propagate that pattern through time and space.

Patterns play two critical roles in Natural Selection:

1. **Reliability** – For a solution to be viable, it must reliably produce a consistent outcome. An unreliable solution, one that works inconsistently, is too costly to store and propagate. Thus, a solution found by natural selection must produce regular, reliable outcomes, leveraging environmental patterns to enable the entity to propagate across generations.

2. **Reproducibility** – The solutions found by natural selection must be reproducible. While current evolutionary theory treats this as mere inheritance, for a Patternist, reproducibility means that solutions are patterns themselves.

Natural selection should be called Natural Elimination. Natural selection is fundamentally a filtering mechanism: variations arise by various means, and the environment eliminates those that cannot propagate.

There are no limits to the environment's influence. Time, natural disasters, interactions with other intelligent and adaptive agents, and even competition within an entity's own group all contribute to the elimination process. Existence itself is a selection factor, with the universe acting as the ultimate selector.

This elimination process results in a set of naturally occurring phenomena shaped by the universe's fundamental conditions, where regularities, randomness, and the relentless progression of time coexist—a set of phenomena that can propagate itself and develop means to facilitate their own propagation.

We call this set of phenomena, *"life"*.

Living organisms are patterns that have identified and incorporated the patterns that define their own structure. Their physical existence is a direct reflection of environmental patterns. The body forms, organ structures, behaviors, and responses that consistently occur within a species—the defining features of that species—are patterns shaped and filtered by natural selection.

There is no limit to natural selection. In biology, natural selection is often treated as confined to living organisms and bounded by genetics. Invisible walls are erected when the concept is extended to human beings, and even more so when it involves culture and technology—where these walls become fully magical, isolating humans from other forms of life. Natural selection is treated as though it stops at biology and does not apply to technology, language, entertainment, philosophy, religion, or science—even though its influence is evident in all these fields.

For a Patternist, there is no isolation. All phenomena attributed to human beings exist in the same physical universe and are subject to natural selection. There is only the process of variation and environmental elimination. Human beings are part of the environment, serving as both the entities subjected to elimination and the selection factors that carry out elimination.

All previous work in biology on natural selection—such as kin selection, sexual selection, and artificial selection—still applies but is now generalized across all domains, including technology, culture, and society. The focus shifts to the role of patterns, viewing natural selection as an automatic pattern recognition mechanism.

As a pattern-finding mechanism, natural selection is not without its drawbacks. First, it is difficult to determine externally which patterns an entity has discovered and is utilizing, how effective those patterns are, and over what timescale they operate.

Second, the variations generated by an entity must attempt new solutions to existence while preserving existing ones. The solutions must paradoxically be different, yet the same. The offspring must differ from their parents but still retain core similarities. Variation must occur, but not excessively.

Finally, natural selection depends on the environment to contain regularities. In a stable environment, natural selection can work steadily toward finding effective solutions. In unstable environments, successful solutions may suddenly become nonviable, potentially triggering a massive elimination event—an extinction. Paradoxically, if sudden environmental changes occur frequently enough, they become a regularity that organisms can adapt to.

Everything that exists is subject to natural selection.

In the context of pattern recognition, we will treat natural selection in standard evolutionary terms, while recognizing it as a universal pattern-finding mechanism applied without limits. Its influence extends far beyond genetics, encompassing acts of prototyping that drive technological innovation—whether in commercial products or military applications. It includes the evaluation of scientific theories based on their predictive accuracy, the cultivation of heirloom crops, and the breeding of competitive racehorses. It applies to martial arts matches that identify the most effective techniques or to any individual learning a new skill by testing against reality, reevaluating outcomes, and trying again. It even governs the training of neural networks, where networks are varied, subjected to elimination factors, and refined before further variation. All these processes fall under natural selection.

Vary a population. Subject the population to the environment for selection and differentiation. Those that are not eliminated are regenerated to a new population. Vary the new population and repeat. Through this process, evolution occurs, and patterns are discovered.

Mechanism 4. Categorization

A feature is anything that can be used to separate two entities in a consistent manner. Once entities are separated, they can then be grouped together based on their shared features to form categories. Thus, categorization is the process of separating and grouping entities according to their features.

While features can be used to categorize the reverse is also true: consistent grouping and separation of entities help define features, leading to the identification of patterns. Thus, if we observe consistent grouping or separation of entities, we can infer that a pattern is being followed. Categorization, therefore, serves as a mechanism for detecting the existence of patterns.

A pattern is, paradoxically, both a single entity and a group of entities. When a collection of entities is grouped together by a common set of features, they form a single entity through categorization.

For example, an electron can be viewed as a group of particles sharing the same features, such as a specific mass and charge. If a particle is encountered and measured to have the same properties as those classified as electrons, it is labeled an electron. This principle applies to protons, neutrons, carbon atoms, all species of organisms, and even objects like knives, chairs, or even individual people.

Naming, labeling, identification, recognition, classification, and even mathematical set theory all fall under categorization. These processes involve grouping entities based on feature similarities and differences, whether the entity is an object, event, physical movement, sound, place, state of a system, or even symbols written on paper.

Every word we use, every entry in a dictionary, and even the names of people result from categorization. We categorize subatomic particles, atoms into chemical elements, and molecules into compounds. We break processes into distinct steps to teach and follow. In various arts, we categorize techniques and judge performances as amateur, professional, or world-class.

We categorize the sounds we hear—whether it's a dog barking or a lion roaring. We categorize music, labeling songs as jazz, rock, country, or just plain noise. When encountering a new organism, we try to identify its species. If it's new, we describe its unique features and assign it a scientific name.

On the road, we label objects as cars, trucks, bicycles, police, or accidents. We label politicians as liberal or conservative. We label people we meet as family or strangers. If we reunite with a long-lost friend, we might be surprised by changes in their appearance compared to our memory of them. But upon hearing their voice or observing their mannerisms and personality—the features that haven't changed—we recognize them again and ask questions about the time apart to update our representation of them, ensuring that the name we have stored now matches the new version of our old friend.

Categorization extends beyond language; it appears in basic biological responses. Organisms must categorize objects as edible or inedible, reacting by either eating or rejecting. They categorize sounds as threats or prey, responding by fleeing or stalking. They must distinguish day from night, when to rest or hunt for food.

If there is a differentiation in an organism's response, and the response is regular enough, categorization has occurred. If categorization has occurred, then a pattern has been recognized. This applies not only to multicellular organisms but also to microorganisms and biomolecular processes, like enzymes and receptors. It even extends to machines and mechanisms that sort, like a centrifuge separating fluids by density or the process of distillation.

We categorize items in a warehouse, organize books in a library, and separate animals into their respective enclosures in a zoo. In a supermarket, items are categorized by aisles, shelves, and sections. We have the cleaning supplies aisle, the beverage aisle, and frozen section. If we want an *"apple,"* we go to the produce section, look for the *"apple"* sign, and find apples labeled with stickers carrying unique barcodes. Why is this done? Why can't items just be arranged and labeled randomly? What are the benefits of categorization, and how is the process carried out?

III. The Search for Patterns

Categorization generally involves three steps:

1. **Find a Collection of Entities** – Any grouping of entities can be categorized.

2. **Apply a Sorting Mechanism** – Entities are sorted into groups based on similarities and differences in features, creating categories.

3. **Test for Pattern Consistency** – Use *Differentiation of Outcomes* to confirm that categories represent actual patterns rather than random groupings.

The collection of entities can be anything that can be divided into groups. It can be a list of songs categorized by genre or a set of images that an algorithm must analyze to determine whether the images are cats, dogs, or humans. It can include all the different cities in a given country or the fundamental forces interacting in the universe. It can be a physical location where entities are grouped by their proximity or an actual physical connection, such as the naming of internal organs and muscle groupings. It can be the various invertebrates found in a square mile of forest, later categorized based on their morphology into groups like spiders, beetles, and butterflies.

Categories, once formed, can themselves become collections of entities. More specific and detailed subcategories can then be created. For instance, we can classify different breeds of cats and dogs, or group human beings based on gender, age, or nationality. Spiders can be categorized into specific species. The extent to which entities can be categorized and subdivided depends on the consistency of the features used to define the new groupings.

For example, in taxonomy, organisms are categorized into domains, kingdoms, phyla, classes, orders, families, genera, and species, with each subcategory being increasingly specific in its defining features.

There is no need for a first mover to provide a collection of entities to be categorized. An agent—the entity that carries out categorization—naturally encounters collections of entities simply through interaction with the world. If the agent has a memory system to store representations of entities and their features, and a recognition system to compare features, it can independently perform the act of categorization.

There is no isolation. A collection of entities does not need to be found, sorted, and categorized by a single individual. Instead, a population of agents can explore and categorize the entire collection, with each agent accessing only a small part and categorizing what it can locally reach. The findings of each agent can then be communicated to others through written, spoken, chemical, or genetic means. This process is evident in science, where fields such as biology, astronomy, and geology collaborate to categorize the diverse entities of the natural world. In evolution, it is the entire population of agents—the collective genetic pool of all members of a species—that participates in the categorization process.

The items in a category should ideally share a common set of features by definition. They should follow a pattern if the newly created category is to be of any use. For example, if we label a set of entities as "*arachnids*," we expect all entities within the "*arachnids*" category to share the features of being arthropods, having eight legs, and possessing a set of shared genes.

But how does this occur? With a collection of entities in hand, what are the various ways to sort and organized entities into sets to create categories?

In general, there are four ways in which categories are formed:

1. **Natural Selection** – Random groupings are subjected to differential survival and elimination. Groupings that help an agent survive reflect actual patterns and are passed on through natural selection. This can be viewed as the naturally occurring categories through which the universe filters out ineffective groupings. Every species of organism is a naturally formed category.

III. The Search for Patterns

2. **Prelabeled Categorization** – Entities are categorized and labeled in advance by an external agent. Pre-labeling is often used for training agents, such as teaching machine learning algorithms to recognize categories like "*cat*" or "*dog*" using pre-labeled images.

3. **Recognition** – Agents compare features among entities to form categories. Entities with similar features are grouped together, while those with different features are separated.

4. **Random** – Entities can be grouped arbitrarily with randomness itself becoming a defining feature.

The detection of consistent categories means the detection of a pattern being followed. We can determine if a category is consistent by applying *Differentiations of Outcomes*, which consist of two components: *Differentiation* and *Outcome*.

- **Differentiation** – The category must have defining features that consistently distinguish it from other categories. For instance, the category "*scorpions*" has unique characteristics that set it apart from the category "*spiders*." Similarly, labels like "*chimpanzee*" differ from "*orangutan*" and "*bonobo*." As long as entities are consistently differentiated, categorization has occurred—even if the category is new and requires a new name.

- **Outcome** – The results of categorization should be consistent. When a test is applied to entities within a category, the outcomes should be uniform for all entities in that category and distinct from the outcomes for entities in other categories. Tests can vary widely, including genetic testing, communication, buoyancy tests, competition, survival, or identification tests where a learning agent must label an image accurately. The specific test depends on the categorization's purpose, with the core requirements being reproducibility and predictability.

If categories are consistent, they reflect an underlying pattern, even if that pattern is initially unknown. This approach has driven the development of many scientific theories.

Entities were categorized as living or non-living long before the principles of evolution were discovered. Organisms were grouped by morphological traits before the discovery of genes. Matter was categorized by properties, leading to the periodic table. Electromagnetic phenomena were identified before Maxwell's equations described them. Every experiment ever conducted can be seen as categorizing data to uncover patterns.

We can apply this same approach to uncover the patterns underlying the category we call *"intelligence."* The first step is to ask: "Do we have an accepted set of entities that we can label under the category of *intelligence*?"

The answer is *"yes."* Such a category already exists. The very existence of the word *"intelligence"* means we have identified a consistent set of entities constituting the *"intelligence"* phenomenon.

However, for a Patternist, the traditionally accepted set—restricted to human beings and selected human activities—is inadequate. It is far too limited.

For a Patternist, the entities labeled as *"intelligence"* must be expanded to include all living organisms. The initial feature that defines the category of *intelligence* is: "Any regularly observed activity or behavior displayed by any organism, including the organism itself."

It is then our task, as Patternists, to uncover the set of patterns underlying everything within this expanded category of *"intelligence."*

III. The Search for Patterns

Universal Tools

Since a solution to a problem must be both reliable and reproducible, solutions are inherently patterns. Thus, *Repetition*, *Prediction*, *Natural Selection*, and *Categorization* are not merely mechanisms for identifying patterns—they are universal problem-solving tools. When faced with a difficult problem, we can creatively apply these tools to find solutions.

We begin with repetition. First, we test to ensure the problem is consistent and not random. We adjust and retest to confirm that our potential solution yields consistent outcomes, checking whether our actions affect the problem reliably. We troubleshoot, search for regularities, and identify features—looking for ways to detect hidden features. We study similar phenomena to replicate successful solutions or avoid actions that could lead to similar problems. We make predictions about the problem and experiment to test whether those predictions are correct.

We apply natural selection to the problem: generating variations, testing them, selecting the best result, and repeating the process. Finally, we use categorization, gathering a collection of entities and organizing them into groups based on defining features. We assign names, labels, or terms to record in a dictionary and then examine each group more closely to identify additional features and uncover further patterns.

These universal tools can explore any phenomenon in the universe. They encapsulate the methods scientists, engineers, and even biological evolution use to solve problems.

Since patterns are at the core of *Repetition*, *Prediction*, *Natural Selection*, and *Categorization*, we can seamlessly switch between tools to gain new insights into whatever we are studying. For instance, categorization can be viewed through the lens of prediction: if we identify an entity as a bird, we predict it can fly based on our prior understanding of birds. Similarly, natural selection can be seen as predictive, where an organism's genes represent predictions about its current and future environment—survival indicates accurate predictions, while failure signals inaccuracy.

Conversely, categorization can also reflect natural selection, where random groupings are refined through an evolutionary process, as seen in language development and the creation of words. Repetition leads to prediction, and repeated features lead to categorization. Even natural selection relies on repetition—a stable environment—for evolution to proceed.

Through Patternism and our understanding of patterns, we can approach any problem from multiple perspectives rather than being confined to a single viewpoint.

The ultimate goal for scientists and philosophers is to address the questions of intelligence and consciousness. By applying universal tools and understanding patterns, we can analyze these phenomena and uncover their underlying principles—the patterns governing intelligence and consciousness—and ultimately solve them.

For a Patternist, intelligence is the ability to find and use patterns. Consciousness is the ability to find patterns within oneself and break free from the patterns one inhabits—whether in behavior, identity, or thought. Conscious systems can be built by first creating intelligent systems and then *feeding* those systems back into themselves.

We approach the problem of intelligence through *Repetition*, *Prediction*, *Natural Selection*, and *Categorization*.

We ask: "Can the system we are building detect regularities?" Can it make predictions and calculate their accuracy? Can it run simulations? Can it apply an evolutionary algorithm generally? Can it categorize: detect, compare, and store features? Can it find patterns and uses for those patterns?

Once we have built such an intelligent system, we can apply the same approach to develop a conscious system.

We ask: "Can the conscious system detect regularities within itself?" Can it recognize when it is looping, stuck, or repeating actions without purpose? Can it predict its own actions and choose to act differently— or even do the opposite and be unpredictable? Can it model itself and simulate its interactions with the environment? Can it categorize itself, labeling events, objects, actions, and images as either "*self*" or "*not-self*"? Can it accurately determine what constitutes its body? Can it choose, control, and change the pattern that defines itself? Can it self-evolve?

A conscious system will inevitably have a body—whether flesh and blood or digital signals on a server. This body will include sensors to perceive patterns and a medium to store the representations of learned patterns. Natural selection can shape the system's physical form, determining what constitutes the body and where to place pain sensors to alert the system to damage—just as evolution has shaped the body forms of innumerable living organisms.

Finally, we can combine all four tools—*Repetition*, *Prediction*, *Natural Selection*, and *Categorization*—to work together or even compete within a unified intelligent and conscious system.

The Divide

A pattern is a discernible regularity. Recognition is the detection, comparison, and storage of features. In pattern recognition, we aim to identify the features of a pattern and store them in a medium—whether neurons, genes, a scientific model, written symbols, or language.

But there exists a fundamental separation between the pattern and its recognition.

Although we describe patterns as being composed of features, this is not entirely accurate. When a Patternist refers to patterns as having features, we are speaking about the recognition of the pattern, not the pattern itself. There is a gap—the pattern is unattainable.

It can be confusing and paradoxical, yet there is an undeniable gap between reality and its representations. It is when we mistakenly believe that we have seized reality in our hands and forget that we are merely holding onto a representation that things go wrong.

For a Patternist, there is an immense divide between a pattern and its recognition. We can only store representations of a pattern, not the pattern itself. All we have access to is the recognition, not the pattern itself. We must always be aware of this divide for within it lies the potential for deception, illusions, hallucination, and bias. Even camouflage and the way drugs trick receptors exploit this gap between a pattern and its recognition.

It is a trap. The result of the trap is the category of phenomena that is the misidentification of patterns. For a Patternist, phenomena such as deception, illusions, hallucination, bias, camouflage, the effects of illicit drugs, and even the manipulation of language and propaganda are genuine biological phenomena, and thus must be included in a *complete theory of intelligence*.

All the tools we use to find patterns—*Repetition*, *Prediction*, *Natural Selection*, and *Categorization*—have inherent weaknesses that can lead to the misidentification of patterns.

Improper use of *repetition* can lead a learning agent to develop biases. Through repetition, we identify patterns by examining a collection of entities and noting what repeats across the entire set. Ideally, this collection should be comprehensive, covering all related phenomena.

For example, to identify all tree species in a forest, we must examine every tree. To understand human regularities, we would need to observe every person on Earth. However, analyzing every entity is often infeasible due to the sheer amount of data. In practice, we study a subset and assume what holds for the subset applies to the entire collection.

If the subset is not large, random, or representative enough, our pattern will be biased. We end up making assumptions about the entire collection when, in reality, our findings may only apply to the subset. Categorization can produce similar biases when based on limited or unrepresentative samples.

The weakness of *prediction* lies in the unpredictability of the universe. The world is messy, with countless variables—both known and unknown—that affect events and outcomes. In a controlled lab setting, variables can be limited, but in the real world, accurate predictions become far more difficult, especially for complex phenomena. Additionally, there is the question of whether an inaccurate prediction is better than none. Does the potential reward of betting on low-probability odds outweigh the cost?

There is a spectrum of predictability. Some phenomena, such as those in physics and chemistry, are highly predictable. Others, like weather, vary in accuracy. Biological phenomena, with their intricate interactions, offer limited predictability. Finally, there are humans, whose consciousness allows for deliberate unpredictability—especially when we realize we are following a pattern.

Prediction also relies on accurate feature comparison, which depends on properly functioning sensors, such as eyes and ears. As these sensors are representations themselves, they are subject to *The Divide*, limiting their reliability and making them susceptible to manipulation, as seen in magic tricks exploiting sensory assumptions.

Natural selection also faces challenges with *The Divide*. First, we do not fully understand the patterns exploited in natural selection. *The Divide* exists between hereditary material—genes—and their functional expression. Bridging genetic sequences with proteins and their specific roles within organisms involves many complex steps.

Second, variation is essential. For natural selection to occur, genetic mutations introduce heritable differences. However, this creates a paradox: genes must simultaneously serve as the recognition of a stable pattern and also as variations to test whether the pattern has changed. Moreover, we cannot distinguish between stability and variation until the outcome is clear—until it becomes evident whether the organism survives and reproduces.

Entities can be *miscategorized*. Ambiguity arises when an entity shares features with two different categories. For instance, should tomatoes be classified as fruits or vegetables? Should the Tasmanian wolf be grouped with the American wolf, which shares its carnivore body design, or with the kangaroo, to which it is more genetically similar? This represents a conflict of patterns, and the decision on which pattern to use depends on the agent performing the categorization process.

Mis-categorization can also be deliberate. The divide between a pattern and its recognition is itself a regularity and can be exploited by any intelligent agent.

The misuse of categorization leads to propaganda and language manipulation. Every word in a dictionary can be viewed as a category with defining features. Manipulating language involves finding words that meet primary features while carrying additional connotations—positive or negative—depending on the desired message.

For example, political events might be categorized as either *"riots"* or *"protests."* Events aligned with the main narrative may be labeled protests, while those opposing it are labeled riots. Whether participants are celebrated or penalized depends on this categorization. The agent setting the narrative determines the label: one state might call an event a riot, while another might label it a protest.

III. The Search for Patterns

The misuse of language is abundant among human beings because we recognize *The Divide* and fully exploit it. We distort the meaning of words until they no longer match their meanings, until the recognition no longer matches the pattern. A break is inevitable. Words either evolve, taking on new meanings, or those who manipulate the language collapse, allowing the language to be restored.

A pattern may not always remain a pattern. It must be constantly checked to ensure it remains a reliable regularity. Similarly, representations must be verified to ensure they align with what they aim to represent. Words must match their meaning, and the recognition of a pattern must always be checked to confirm that the defining features triggering recognition remain true.

A lapse in vigilance can lead to failure, blindness, or even extinction. A Patternist must remain aware that what we believe to be true might not be. What we hear and see may be a lie. A scientific model can be wrong.

We must constantly check and recheck, for patterns can change, they can evolve, and they can suddenly cease to be.

IV. A Theory of Theories

What is a Theory?

A theory is a representation of real-world phenomena. Ideally, a theory should be able to:

1. **Predict** – Make predictions about phenomena, such as event outcomes or entities to search for in reality. These predictions rely on defining and matching features.

2. **Simulate** – Create models of phenomena using physical models, computer simulations, or language. Simulations can focus on the past, present, future, or customized scenarios, generating observations that lead to predictions about reality.

3. **Organize Information** – Transform raw data by categorizing and labeling it. A theory should consistently differentiate between distinct phenomena and identify when the same phenomenon is observed.

4. **Generalize** – Reduce a phenomenon to a set of fundamental principles. These principles can then be applied across domains and mediums to build, engineer, or solve problems.

5. **Define Governing Rules** – Identify and describe the rules governing the system that allow the phenomenon to occur. This helps define the limits of what is possible, guiding the allocation of resources.

6. **Store, Reproduce, and Communicate Knowledge** – A theory should be documentable, teachable, reproducible, and easily shared.

There is no need for confusing philosophical formulations. The practical uses listed define what makes a theory. If a theory accomplishes these tasks, it is valid, regardless of any philosophical objections. If it cannot, it is ineffective and does not qualify as a theory.

We have shown that predictions are made by observing phenomena for patterns and then using those patterns to calculate predictions. Simulations are created by feeding predictions back into itself to produce the next step in a recursive loop. To organize means to categorize. To generalize is to reduce a phenomenon to fundamental patterns. To define the governing rules means to identify system regularities that are repeatedly enforced. Storing, reproducing, and communicating knowledge rely on patterns. At the core of all practical uses of a theory, we find patterns.

For a Patternist, a theory is the full recognition of the fundamental patterns constituting a phenomenon. A theory should seek to identify all patterns and regularities a phenomenon exhibits, including the phenomenon itself. Our approach to a theory of intelligence will follow these same principles.

For a Patternist, intelligence is the ability to find and utilize patterns. Through Patternism, we predict that the direction of evolving complexity in any intelligent system will be towards faster and more effective means of finding and utilizing patterns. Since adaptation is intelligence, we predict that any adaptive system will consist of a *randomizer*, a *sorter*, and a *flexible medium* using the mechanisms of *Repetition*, *Prediction*, *Natural Selection*, and *Categorization* to find patterns.

We will build simulations of intelligent systems. In fact, simulations already exist in machine learning—employing evolutionary algorithms and neural networks to recognize images, generate text, translate languages, drive cars, and detect patterns in data. However, what remains missing is a comprehensive explanation of why these programs work. Scientists have replicated the biological phenomenon of neural networks in software, often treating them as a black box without fully understanding the underlying mechanisms for their functionality.

IV. A Theory of Theories

Through Patternism, by understanding patterns in terms of features and regularities, we can fully explain how and why these programs work, starting from first principles as well as explain how it is possible that biological neurons can be translated into algorithms and result in intelligent behavior.

Through Patternism, we will examine the genetic algorithm, the linguistic algorithm, and the neuronal algorithm, discovering they all share an underlying structure—*Universal Recognition*—a single, unified algorithm that operates on patterns and is present in all intelligent systems.

Through Patternism, we will redefine the phenomenon of intelligence. We will organize and expand the category of intelligence to include all living organisms and their behaviors, including all human activity and behavior. Our approach will surpass the achievements of evolutionary theory, creating a knowledge base that includes not only biological, but also manufacturing and technological phenomena, as well as insights into consciousness, communication, and science.

What a Patternist seeks in a theory of intelligence is to find and describe its fundamental *principles*. It is strange. To develop theories like Newtonian mechanics, the periodic table, or Darwin's natural selection, scientists observe regularities in phenomena and generalize them. Yet, when it comes to intelligence, no such process occurs.

It is a seeming blindness. All those who claim to be searching for a theory of intelligence have chosen to wander aimlessly instead of using the proven methods that have produced so many accurate and practical theories.

Through Patternism, we have the means to develop a theory of intelligence. We will examine all biological phenomena and use them as clues, recognizing that the intelligent and biological phenomena are one and the same. What we will uncover are the principles of *Representation*, *Recognition*, *Reproduction*, and *Randomization*—a *viable* set of interacting principles underlying all intelligent and biological systems.

We will identify the governing rules not only for intelligence but for the universe itself. Patternism is the belief that patterns—and the *Paradox of the Patterns*—are fundamental aspects of the universe, in the same manner as space, time, and matter.

We will show that knowledge storage occurs through representations, that reproduction is the propagation of patterns, and that communication, in all its forms, relies on patterns. This book itself serves as a medium to store, reproduce, and communicate *a complete theory of intelligence* grounded on the first principle of patterns.

Finally, we will account for the process of theory creation itself, showing those interested in the phenomenon of intelligence how to craft their own theory, find their own set of principles, and refine them—if what this book has to offer is deemed insufficient.

IV. A Theory of Theories

Foundational Principles

All theories are built from a foundational set of principles.

All of the great scientific theories that have produced accurate predictions and engineering feats have taken real-world phenomena and broken them down into a working set of principles.

Newton's three laws of motion form a foundational theory of motion. Einstein's theory of special relativity rests on the principle that the speed of light is constant for all observers. The theory of evolution relies on the interactions of variation, heritability, differential survival, and reproductive success. Atomic theory is built on the principle that all matter is composed of indivisible units. Kepler's laws of planetary motion. Maxwell's equations for electromagnetism. Even the adaptive immune system can be said to operate unconsciously on the principle that pathogens can be identified by their unique protein receptors.

A scientific theory is defined by the principles it proposes. When these principles accurately reflect the regularities found in nature, the theory will account for all related phenomena, enabling accurate predictions, simulations, and engineering feats. If the principles are incorrect, the theory will fail to explain phenomena, rendering all its predictions invalid. A theory with partially accurate principles may still be useful, though incomplete, and can serve until a more fundamental set of principles is discovered. The worst-case scenario occurs when incorrect principles are mistakenly accepted as true, leading to decades of wasted time, resources, and delays in uncovering a valid set of principles.

The same holds true for mathematical theories. All branches of mathematics—from graph theory, set theory, and number theory to Euclidean geometry—are built on their own set of principles, specifically called axioms. From these axioms, theorems are derived, proofs are constructed, and the theory is systematically developed.

In mathematics, everything stems from the set of axioms. All allowable manipulations performed on a sequence of symbols must be based on the axioms. The result of such strict adherence is that, given the same mathematical theory and axioms, a mathematician working on Earth would derive the same outcome as a mathematician working on Mars or anywhere else in the universe. When axioms are not strictly followed, errors and inconsistencies arise.

Mathematics follows patterns, just as the universe follows patterns. This unspoken, inextricable link between mathematics and the physical world is why math "*works*." Strict adherence to a set of axioms mirrors the way fundamental aspects of the universe adhere to physical laws. All mathematics can be reduced to the consistent combining and separating of symbols and understood as the interaction of strict patterns. This interaction is what gives mathematics—mere symbols written on paper—its predictive power. Furthermore, any system that follows a strict set of axioms, patterns, or governing rules defining how entities within that system are separated or grouped can be freely translated from one system to another. This allows mathematics to model any phenomena in the universe, provided the phenomena exhibit strict regularity.

We can also view mathematics from the opposite direction, where we are not initially given the axioms but must deduce them. Given all our observations of a mathematical theory's steps—all derivations, proofs, and logical progressions—can we determine the underlying set of axioms? How would we proceed, and has this been done before?

Can we extend this approach to the universe itself? Suppose the universe operates according to an unwritten set of axioms. Would it not then be up to us to uncover these fundamental axioms through our observation of everything that exists within it?

Is this not the transcendent view shared by mathematicians? Is this not the objective of all physicists conducting experiments? Is this not the very task of philosophers and scientists?

We can view foundational principles in terms of physical objects and their fundamental component parts. For example, we can examine all biological proteins and ask, "What are the fundamental components of all proteins?" Through experimentation, we find that all biological proteins are composed of 21 amino acids. Similarly, we can ask the same question of genes, discovering their building blocks:

IV. A Theory of Theories

four nucleotides, codons, and the genetic code. Applying the same logic to machines, we uncover engines, screws, gears, and mechanical principles such as levers and pulleys, along with laws like Archimedes' and Bernoulli's principles. Likewise, in electronics, we break down systems into diodes, resistors, capacitors, inductors, power sources, and so on.

We do not reduce for the sake of reducing. We reduce to build and create. Once we identify the fundamental components, we can use them to design entirely new entities. We create new proteins. We engineer new genes and insert them into organisms, causing them to display features they have never shown before. We build new machines and electronic circuits by combining basic components in innovative ways. Furthermore, since all of these objects exist in the same physical universe, we can freely apply principles from one domain to another, combining components or concepts originally developed in different fields.

The world is complex. Extremely complex. Yet it must be represented and stored within the human brain—within networks of neurons—using a form of representation with predictive power that allows its user to navigate the world. This process gives rise to a variety of theories, each built upon foundational principles. Philosophical, political, and economic theories—all the various *-isms* and doctrines—are structured around core principles that define not only the values shared by their adherents but also how they view and interact with the world.

Thus, we have the *-isms*: Secularism, Liberalism, Communism, Socialism, Capitalism, Realism, Aristotelianism, Platonism, Stoicism, Machiavellianism, and countless others. Each has its adherents—the *-ists* and *-ians*—and is subject to evolutionary forces that either seek to conserve the theory as it is, or transform it. Over time, these changes can become so radical that a new theory—a new *-ism*—emerges, branching off from the original and evolving independently, much like species branching out and evolving into new forms over time.

There are no limitations.

Our theory of theories applies to government. A theory about how a government should ideally function has its foundational principles stated in its constitution. The foundational principle of a monarchy lies in the family name of the ruling king.

Our theory of theories applies to religion. Buddhism has the Four Noble Truths and the Eightfold Path. Christianity has the Ten Commandments and the Bible. Islam has the Quran. All religions have a set of tenets, a set of beliefs, that provide structure for how adherents should live their lives.

What is the truth in a mathematical or scientific theory? What is the truth within religion and a book of teachings? What is the truth within a sequence of DNA?

Whether these theories are *true* or not, whether the *-isms*, doctrine, or constitutions reflect reality or simulate it with high accuracy, does not matter to a Patternist.

To a Patternists, **all theories are patterns**. Each theory can be viewed as a single pattern, with its features being its foundational principles. To a Patternist, the central question is whether a pattern can propagate itself through time and space. The only thing that matters for a theory—regardless of what it applies to—is whether it can sustain itself and endure across time and space.

There is no isolation. Theories do not exist in the abstract—they exist within physical reality. A theory's representative token is language. Its storage medium is language. Its propagation occurs through human brains. Human beings serve as vessels for theories.

Do not take a theory and judge it in isolation. Theories must be examined in totality, along with the vessels that carry them. If a theory reflects reality and proves useful as a tool for predicting outcomes, then so be it. If it spreads virally through populations, latching onto human nature, emotion, and desires for power, then so be it. If it helps an individual, a tribe, or a civilization maintain existence, using them as its vessel and sustaining that vessel across generations—against natural disasters, decay, and wars waged by other tribes, civilizations, or competing theories—then so be it.

IV. A Theory of Theories

And if it goes extinct, surviving only as a forgotten book in a library, a fossil of what once was or what could have been—if it lives and dies within the mind of a single human being—then so be it.

Constructing Reality

From a set of principles, a scientific theory is formulated. From a thesis, a book is written. From a set of axioms, mathematics is derived. From a set of beliefs, a religion is established. From a set of physical laws, a universe is created.

There is an underlying structure to all forms of reality.

Principles, axioms, beliefs, and laws are all variations of the same fundamental phenomenon: regularity—the essence of what it means to be a pattern.

Our perception is built from patterns. We construct a reality from the regularities we observe. This applies not only to all theories, but to all forms of simulations, including entertainment.

Entire worlds are created.

Simulations are built by taking an initial moment in time and feeding it into a *Prediction Engine*. The *Prediction Engine* uses the features of that moment, along with learned patterns, to calculate the next step—the next moment in time. This next moment is then fed back into the engine to calculate the following step, and so on, recursively creating a simulation.

When we program this process into a computer, we create simulations that generate entire worlds from a foundational set of code. This gives us not only physics engines, computer-aided manufacturing tools, and weather forecasting systems but also video games. Each game, running on its own set of code and engine, presents a new world for players to inhabit and explore.

The phenomenon of video games demonstrates how entire realities are created and experienced. We have the *Mario* world, the *Tetris* world, and the *Elder Scrolls* world, each with its own distinct set of rules and interactions.

The creation of video game worlds mirrors our understanding of evolution and patterns. Examining the *Mario* world and its various iterations reveals it as a collection of worlds, each built on a different game engine. The *Mario* world is a franchise; there is no single *Mario* game. The *Mario* world is a *Mario* universe, featuring recurring iconic characters, gameplay styles, and lore. Since there are iterations of *Mario* worlds, each following the next, we can say the *Mario* universe evolves over time. A *Mario* game from 1990 is drastically different in both visuals and gameplay from one released in the 2000s, 2010s, or 2020s. And yet, they remain the same, still labeled as *Mario* games because we can identify the features that define the *Mario* pattern.

There is no isolation. The fundamental structure underlying all video games also applies to all theories. Whatever questions we ask about a scientific or mathematical theory, we can also ask about video games—and vice versa. The insights gained from one domain can apply across all.

We can generalize reality.

We can ask: "Is a video game consistent?" Does it run smoothly, or does it crash? Are there bugs and errors? What differentiates version 1.1 from version 1.2? How is the game engine created and implemented? Can players exploit the code? Answering these questions deepens our understanding of all representations, including our own reality.

When we play a video game, we enter a reality governed by its own set of fundamental laws. Switching to a different game means stepping into an entirely new reality, where we must learn new patterns to understand how the new reality operates. But why stop there? We can explore entirely different universes altogether.

We can freely hop between the various realities presented as entertainment. We enter the world of fiction—movies, comic books, TV shows—where each story showcases a universe with its own unique set of regularities and governing rules. There is the *Marvel* universe, the *Harry Potter* universe, the *Lord of the Rings* universe, the *Star Wars* and *Star Trek* universes. There is a *Vonnegut*, *Hemingway*, *Homer*, and *Ovid* universe.

We can even visit different versions of reality that aim to describe our own: histories according to the United States, Imperial Japan, or the Soviet Union. History from the left, history from the right, and history according to *Herodotus*. The heroes in one account may very well be villains in another. To believe any one representation is the ultimate truth is to fall into the trap of *The Divide*—the gap between a representation and what it seeks to represent. Yet to disregard these accounts entirely would be equally foolish, for there may be hidden value within.

We strive for consistency in our reality, theories, and the worlds we explore.

We don't want a theory that predicts one outcome while experiments yield another. We don't want our mathematics to produce two different answers when the axioms have been followed exactly. We don't want to see sneakers in a film set in ancient Rome. We don't want a character in a story we're engrossed in to behave *"out of character."* And if they do, we want to be able to look back, reread, or rewatch and realize the author gave us plenty of hints—making it our own fault for missing them.

We want consistency.

But what does it mean to be consistent?

For a Patternist, consistency means regularity—it means following patterns. When our theories and reality are consistent, we can predict, run simulations, engineer, and achieve practical outcomes.

We want principles, axioms, and rules to be applied equally, to all, and at all times. We don't want them followed one day and broken the next. We don't want new rules to suddenly appear one day and vanish the next. If the underlying rules are constant, we will eventually figure them out, even if by brute force. But if the rules keep changing, we can never understand them, nor would it be worth trying. By the time we describe and apply the rules, they will have become useless.

It is confusing to get a different result in a theory, in math, in a computer simulation, or in a chess game every time—even when the exact same variables, steps, conditions, and methods were followed.

So, we strive to make our reality and representations consistent. We strive to live up to the principles we hold to be true.

We do our best.

When an inconsistency arises in our reality, theory, or beliefs, we talk about it. Whether the inconsistency occurs on a personal, societal,

IV. A Theory of Theories

or universal level, we point it out to draw attention to it, hoping to resolve it or warn others of a potential trap. Sometimes, we laugh at it to cope, easing the stress and frustration of our inability to resolve the inconsistency. When others fail to live up to their own stated principles—the same ones they expect others to follow—we call it hypocrisy.

We set up systems to resolve the inconsistency.

In science, communities discuss theories, proposing and performing experiments to determine a theory's consistency with itself and with nature. Maps, computer programs, software, and hardware are constantly updated to newer versions after fixing old bugs and exploits. *George Lucas* once decided what was canon in the *Star Wars* universe, but when he sold *Star Wars* to *Disney*, the company took over the role of officially deciding canon and resolving inconsistencies among the various storylines written by different authors. In religion, we look to clergy to interpret and apply doctrines.

The extent to which we try to keep our theories and realities consistent is total. All governments engage in this effort. Lawmakers, lawyers, judges, and the police all play a role. The *United States* has its own theory of law and justice, outlining the rules citizens must obey and the consequences if those rules are broken within its jurisdiction. The principles that underpin this theory are officially stated in the *U.S. Constitution*. To remain consistent, all laws passed, as well as how trials and punishments are carried out, must adhere to the *Constitution*, the *Bill of Rights*, and applicable state laws.

When inconsistencies arise—when laws conflict with each other or with the principles outlined in the *Constitution*—we rely on the courts, ultimately the Supreme Court, to interpret and resolve these issues. If we remain unsatisfied, we have the power to amend the *Constitution*, adding new principles or eliminating outdated ones to ensure the government of the *United States* better aligns with an even greater, yet unwritten, set of principles of freedom, justice, and maintenance of the existence that is the *United States*.

We do our best to maintain consistency.

But the world is a messy place.

The various systems and institutions we establish to maintain consistency may become corrupt and fail in their duties. Our principles, axioms, theories, ideals of our country, traditions, history, perception of human nature, the people we admire, the people we hate, the laws we obey—everything we believe to be true—may turn out to be a lie. Our world, our reality, can come crashing down.

What can we do when it inevitably does?

The world is such a messy place.

Should we continue to hold on and lie to ourselves? Continue to live in our poorly constructed reality? Ignore the cracks in our foundations? Maybe it is for the better. The divide between our reality and ultimate reality seems insurmountable at times.

And what if it's just a small lie? The lies that a society perpetuates may lead to its downfall when the flood comes and the walls break beneath the waves. But what about the small lie lived by a single man? Would it really be so costly for a single man to live his lie?

A Patternist will not lie to himself.

A Patternist will not ignore any cracks.

The foundation upon which the human-centric theory of intelligence is built is flawed. It is cracked beyond repair.

The inconsistencies are glaring.

If only human beings are intelligent, then how could evolution—a stupid process—produce human beings, the human brain, and human consciousness?

If all living organisms and their activities are considered part of the biological phenomena, then why don't we treat science, technology, music, computer simulations, video games, fiction, history, religion, and propaganda—all human activities—as part of the biological phenomena, subject to the same fundamental principles?

And why are we not searching for this set of fundamental principles?

The inconsistencies are glaring.

As Patternists, we will not tolerate blindness. We will demolish the false foundation of the human-centric view of intelligence.

IV. A Theory of Theories

Our house of intelligence will be built upon the foundation of patterns. We will treat all living organisms and their activities as part of the intelligent phenomena. We will search for the fundamental set of principles that all biological phenomena share.

We will erect the pillars of *Repetition*, *Prediction*, *Natural Selection*, and *Categorization*. We will erect the walls of *Representation*, *Recognition*, *Reproduction*, and *Randomization*.

We will actively seek out cracks. If we find any, we will take a hammer to them ourselves, tear down the pillars, tear down the walls, tear into the foundation with our own hands—and rebuild anew.

If we find a better pillar, a stronger wall, we will add them.

For a Patternist, the foundation of intelligence must be of the highest grade. We challenge all others to hold themselves to a higher standard than what a Patternist demands of himself: to explain all biological phenomena—including all human phenomena and even theory creation itself.

But there is a single crack in the foundation of Patternism.

As Patternists, we will not lie. We will fully acknowledge it.

That crack, that inconsistency, is the *Paradox of the Pattern*.

For things to be different, and yet the same. For the same things to be different.

All cats are the same, and all cats are different. All human beings are the same, and all human beings are different. All theories are the same, and all theories are different.

I am. I am a different person now than I was yesterday, a year ago, decades ago, and yet I am the same person now as I was yesterday, a year ago, decades ago.

I will be. I will be a different person in the next moment, tomorrow, years from now, and yet I will be the same person in the next moment, tomorrow, years from now.

I am, and I am not. I was, and I was not. I will be, and I will be not.

This inconsistency is absolute.

This crack is undeniable.

But strangely, it does not seem to weaken the foundation. Rather, it strengthens it. The entire foundation is strengthened by the paradox.

Skyscrapers can be built. The crack holds. It is central. When one peers into it, we find a seemingly endless abyss—an infinite core. From it, the phenomena of reproduction, eating, and repair emerge. From it: representation, perception, and deception. From it, self-reference and consciousness. From it, evolution. From it, so much more.

All inconsistencies reduced to a single point, to a single crack.

Bizarre. Absolutely bizarre. We Patternists freely offers all a hammer to strike the paradox and crumble the foundation of Patternism if they can. But it appears to be unbreakable—a seeming void that offers glimpses into the slight glimmers of ultimate reality.

And so, it will not be hidden. It will be on full display. The *Paradox of the Patterns* will be the central core. It will remain the single void, the single inconsistency, the one crack upon the foundation of Patternism.

IV. A Theory of Theories

The Universal Method

Collect a set of phenomena. Observe the phenomena for patterns. Use the patterns as principles for a theory. It was through this *Universal Method* that Darwin developed the theory of natural selection. By observing various species—from tortoises to the finches of the Galapagos—Darwin identified regularities of variation, heritability, differential survival, and reproductive success, forming the foundation of his theory. Similarly, Alfred Russel Wallace independently reached the same conclusion after years in the Amazon rainforest and the jungles of the Malay Archipelago, observing and collecting beetles, butterflies, and other organisms to sell to private collectors and museums.

Assume universality. Aim for universality. The goal is universality.

All great scientific theories were developed using the *Universal Method*, whether the founders were consciously aware of it or not. It is how Kepler and Newton discovered their laws of motion. It is how supervised learning in artificial intelligence operates. In genetics, regularities in genetic material are used to determine a theory of ancestry, showing how organisms are related. Even the act of finding a fitting function for a set of data points is an application of this approach. When scientists perform experiments, make observations, and plot and fit resulting data, they are applying this method.

There are no limitations.

This method of pattern recognition extends beyond science. It is how Joseph Campbell identified the monomyth, gathering myths from different cultures and times, revealing the pattern of the hero's journey. It is how Aristotle wrote *Politics*, establishing his principles of rule by one, rule by the few, and rule by the many, by observing the different forms of government across the city-states of ancient Greece. It is how Robert Michels discovered the iron law of oligarchy by studying the evolution of democratic organizations.

It is how the Buddha developed his theory of human suffering from his observations of ordinary life outside the walled garden of his palace home. And it is how the Buddha arrived at a method to end individual suffering—by venturing into the world to see, learn, and experience.

And it will be through this method that we will find and describe a *complete theory of intelligence*.

The steps of **The Universal Method** consist of:

1. **Collect a Set of Phenomena** by exploring, experimenting, or observing the environment.

A display case of beetles is a collection of phenomena. Butterflies in an observatory are a collection of phenomena. A stack of object-labeled photos. Grandmaster chess games with recorded move sequences. Recordings of a spoken foreign language with their translated subtitles. The set of all biological organisms, observations of celestial objects in the night sky, or studies of physical motion—all are collections of phenomena.

To produce an effective theory, our collection must be high in both *quality* and *quantity*.

High *quality* ensures that the features of the phenomena are thoroughly recorded without omitting details. Direct access to the phenomena is preferable, as relying on secondary sources, such as textbooks or others' research, risks missing key features essential to our search. Secondary sources may also be outdated or lack the precision of modern observational tools.

We also want a large *quantity* of phenomena in our collection to reveal regularities. A pattern is a discernible regularity, and since repetition is our primary tool for identifying these regularities, a larger collection makes repetitions more apparent. For instance, a repetition that occurs 100 times is easier to detect than one that appears only 3 or 4 times. When a collection contains only one or two phenomena, repetition cannot be relied on, making regularities difficult—if not impossible—to detect, thereby rendering the *Universal Method* ineffective. A small collection is also more likely to introduce bias, potentially leading to a false theory. Ideally, the collection should be comprehensive, encompassing all related phenomena in the universe to ensure that the resulting theory is both general and universally applicable.

Once we have a collection of phenomena and recognize them as what we seek to explain, any theory we develop must cover all phenomena in the collection—no exceptions. We cannot selectively include or exclude phenomena; our aim must be universality, ensuring all data points are connected. If our theory cannot explain a phenomenon, we must address why, acknowledge the theory's limitations, and accept that a more comprehensive theory awaits discovery.

2. **Observe the phenomena for Patterns** by examining each phenomenon in the collection. Use instruments, measurements, and recordings to break each phenomenon down into its features, looking for regularities. The more general, consistent, and fundamental a regularity is, the better.

All the tools of pattern recognition—*Repetition*, *Prediction*, *Natural Selection*, and *Categorization*—apply here, with repetition of features being the most readily detected.

In practice, we deconstruct each phenomenon into a list of features and repeat this process for all phenomena in our collection. We then compare these lists, looking for recurring features. These recurring features are gathered for further examination, and if they are fundamental enough, they serve as principles for a theory.

Unfortunately, theory creation is complex.

If only one feature appeared across all lists, creating a theory would be simple, but this is rarely the case. Instead, we often encounter a set of features, with each phenomenon displaying only a few at a time. For example, we will break down all intelligent and biological phenomena into the features of *Representation*, *Recognition*, *Reproduction*, and *Randomization*. However, if we were to pick a random biological phenomenon, we might find only one of these features prominently displayed.

Such cases may indicate that our theory is not fundamental enough and that what we have found is an intermediate theory. Intermediate theories are still valuable, as they can serve as stepping stones toward a more fundamental theory.

Creating a theory would also be simpler if the list of features into which we've deconstructed a phenomenon were short and complete. But this is rarely the case. Lists can be lengthy or contain only partial information. Complexity deepens when we recognize that features are patterns in themselves, each capable of being broken down further into sub-features, which can then be broken down into sub-sub-features, and so on.

The search for regularities becomes vast and intricate, resulting in a *"list of lists of lists,"* and all we can do is hope that somewhere within these nested lists lies the regularity we are searching for. Moreover, the critical features we seek may not yet be fully described or may not even have a name.

The features that make up the regularity must be detectable. The methods used to examine our collection must be able to capture and record the defining features of the underlying pattern. For example, it would be challenging to develop a theory of diseases without microscopes, to study celestial bodies without telescopes, or to understand species relationships without DNA sequencing.

This principle extends beyond visual or spatial features. Temporal patterns—regularities that unfold over time, including actions, sounds, movements, behaviors, and timing—must also be considered. Therefore, tools that can record, store, and compare temporal features are essential for successfully applying the *Universal Method*.

After the diligent work of collecting and observing, we may ultimately have to rely on a measure of luck to uncover the fundamental regularities needed to craft a theory. It may require deep meditation. Isolation. Endure a bout of malaria-induced fever dreams. Starve ourselves into a state of delirium. From there, with random luck, the fundamental patterns for a theory might bubble up from the depths of our mind, with the seeds of their growth planted during our initial work of observing and collecting.

IV. A Theory of Theories

Alternatively, we can be strictly methodical and organized, as seen in science. We gather data, organize it into tables and graphs, making patterns visually apparent. When the *Universal Method* is applied to quantitative and experimental science, the features become variables. We search for regularities by fitting equations using the variables derived from the data we collect. If the equation fits, we explore further. In this case, repetition occurs in the equation itself, as the results of the experiment consistently align with the predictions made by the fitted equation, once adjusted by the variables.

3. **Use the Patterns as principles for a theory.** We then use the theory to make predictions, program and run simulations, and engineer solutions, thereby testing the accuracy of our theory in the process.

If our theory is accurate and our principles are fundamental, the theory should not only explain all phenomena in our initial collection but also extend beyond, allowing us to explain, predict, and engineer new phenomena. If our theory explains the initial phenomena but cannot predict or engineer new phenomena, we may have introduced bias by using an incomplete or insufficiently large collection of phenomena. In such cases, we may need to expand our collection to ensure it is comprehensive, eliminating selection bias that could lead to false regularities and, ultimately, false theories.

In all cases, finding patterns is beneficial. Even if we cannot construct a theory from the regularities we've identified because they are not fundamental enough, we can still use these regularities as clues in the search for a potential theory. In this context, the regularities simplify matters and serve as target outcomes that a potential theory must explain and predict. For example, the gas laws equation relating temperature, volume, and pressure was discovered experimentally before the kinetic theory of gases was developed.

In practice, the *Universal Method* is implemented through representation in two primary ways.

The Role of Representation in the *Universal Method*:

1. **Features can be represented as words, symbols, or any other representative form.** Since a feature is anything that can separate and group entities in a *consistent* manner, we can define any feature for a given phenomenon as long as the definition remains *consistent*. This requirement for consistency means that features are patterns and can be effectively represented as words, symbols, pictures, neurons, or any other *representative forms*. As long as a *translation machinery* exists to translate the *representative form* back into its *original form*, we can effectively store, share, and compare features using the representation instead of the actual phenomena.

   This gives rise to making descriptions and taking notes—literally using words to describe entities. Descriptive terms like *"black," "blue," "soft,"* and *"hairy"* label features. Additionally, combining words into sentences enables us to describe behavior and interactions that an entity displays, marking and storing these as features of that entity.

   Through feature representation, all methods of recording become possible. Representing features enables drawings of flying frogs to be made and preserved in a biologist's field journal, photos to be taken of microbes, videos to capture mating dances, and audio recordings to document the various ways organisms communicate.

   Representation reduces complex, temporal, or abstract features of a phenomenon to a single word or concept, simplifying the search for regularities. For example, a theory of evolution could not have emerged without pre-existing words like *"reproduction," "inheritance,"* and *"variation"* in the minds of Darwin and Wallace. Similarly, a Patternist theory of intelligence could not exist

IV. A Theory of Theories

without terms like *"representation," "recognition," "randomization,"* and *"pattern"* already defined and in use.

2. **Through representation, the medium in which the *Universal Method* is carried out does not matter.** Whether patterns are detected through neurons with threshold potentials, silicon transistors with machine learning algorithms, symbols on paper, graphs and tables, simulations, or language—makes no difference. As long as the medium can detect repetition and regularities, the *Universal Method* can be applied, and all the tools of pattern recognition can be used within representations to identify patterns.

Furthermore, multiple layers of representation can exist to support the *Universal Method*. In human beings the method is ultimately carried out through networks of neurons, with language, math, and science serving as intermediate layers, each contributing to the process.

The ability to represent features—to capture, store, and share them across different mediums—is crucial. Representation is why libraries exist, allowing students to read books and learn. Scientific journals exist because experimental results can be recorded and shared, allowing a wider community to join in the search for theories. Representation enables a biologist to document and share discoveries from a lost jungle with colleagues on the other side of the globe. It enabled a patent clerk, by reading various experimental attempts to measure the speed of light, to develop a revolutionary theory without conducting the experiments himself.

Through representation, we harness the computational power of entire populations. It is no longer necessary for the individual who collects phenomena to be the one who finds a theory to explain them. They need not be in the same room, or even live in the same century. All that is required is that features are accurately recorded.

The search for fundamental patterns is outsourced. A million brains will outwork a single brain. One viable theory advances human knowledge. This is especially true in the modern age, thanks to technological innovations. We now have sensitive instruments that record finer details, video technology that captures high-definition visuals and slow-motion movements, and digital tools that free us from the limitations of ink and paper, allowing immense data storage. With the internet, we can instantly share all the information we collect with millions of others.

We can use the *Universal Method* to find a theory of intelligence.

The *Universal Method* is viable as long as:

1. **There is a sufficiently large collection of phenomena.**

2. **The underlying pattern's features are detectable.**

We have an overabundance of data. How much information—features and their patterns—has been collected and described about various biological phenomena over the past century? Or just in the past decade alone? How much information has been gathered about human beings alone—our behavior and all the activities we engage in? How many high-quality recordings, experiments, and instances of rigorous documentation using modern equipment and sensors to detect finer features have been conducted by scientists and shared through academic institutions?

We have an abundance of brains to find a theory. How easy is it now to go online, read articles, listen to lectures by philosophers, scientists, and biologists, or watch videos of actual neurons, bacteria, protists, and other living organisms? How easy is it to connect with others, forming communities and assisting one another in the search?

The *Universal Method* is ripe for use.

So why isn't it being used?

Why has it not been applied to find a *complete theory of intelligence*?

IV. A Theory of Theories

The human-centric approach to intelligence is fundamentally wrong and is the greatest obstacle to finding a *complete theory of intelligence*.

All that is needed to use the *Universal Method* is an abundance of phenomena to search for regularities within. But the human-centric approach prevents this. The false belief that only human beings are intelligent severely limits the reach of our search to a single phenomenon. The *Universal Method* cannot be applied. It cannot be used effectively when the collection of phenomena consists of only one.

The human-centric approach is wrong. The human-centric approach has provided key insights into intelligence, but it has reached its limits. We must go beyond.

What happens when we free ourselves from the chains of the human-centric view? What happens when we expand the phenomena of intelligence to include all living organisms and their activities?

We will be working with big data.

We will be overwhelmed with an abundance from which the principles of a *complete theory of intelligence* can be readily drawn.

The work of every scientist, past and present, becomes a clue. How bees communicate. How ants find food. The adaptive immune system. Genetics and sexual reproduction. The active camouflage displayed by chameleons and cuttlefish. Every sensory organ. All the research that took years or decades to document and catalog—once dismissed and ignored by those claiming to seek a theory of intelligence—now plays a critical role.

Every nature documentary becomes a record of the intelligent phenomena. Every living organism and its activity is part of the intelligent phenomenon.

The biological and intelligent phenomena are one and the same.

And we go beyond. Every single human activity is now included in our definition. The printing press, the internet, martial arts, music, video games, sonar and infrared sensors, manufacturing techniques, measurement tools, blueprints, neural networks, math and science. Consciousness itself. All will be included in the phenomena of intelligence.

Patternism

Collect a set of phenomena. Look for patterns within. Use the patterns as principles for a theory. We will use this method—the *Universal Method*—to find the principles for a *complete theory of intelligence*.

V. A Viable Set

Merely Viable

Given the vast observations of all living organisms and their activities—including human beings and human activities—can we find a set of patterns to use as principles for a theory?

The question must be asked again.

Given the immense amount of recorded data about living organisms and their activities—including human beings and all human activities—can we identify a universal set of principles that all living organisms, including human beings, use to survive, thrive, and propagate in the same physical universe, subjected to the same natural disasters, droughts, floods, predation, competition, and decay of time?

Does such a set of fundamental principles exist?

For a Patternist, the answer is *"yes."* This set of principles not only exists, but all efforts must be made to find and describe it.

Yet, why is this not done?

Why are we not searching for and describing this set of fundamental principles?

We have all the resources. We have access to vast amounts of information. We have near-instant communication.

Why, then, is this not done?

All other great scientific theories break the phenomena they seek to explain into a working set of interacting principles. Newton reduced all motion of physical objects into three fundamental principles known as Newton's three laws of motion. Maxwell's equations describe all electromagnetic phenomena. Chemistry breaks down all matter into the bonding of atoms, listed as elements in the periodic table. Quantum mechanics goes further, describing matter through the interactions of elementary particles within the framework of the Standard Model.

Yet, when it comes to intelligence, this is not done. When it comes to the set of all biological phenomena, this is not done.

Do we not live in the same universe as all other living organisms? Are we not subject to the same fundamental laws of physics?

Are human beings isolated by a magical wall—a barrier erected by ego and arrogance that claims only human beings are intelligent? All other organisms are stupid. Viruses and bacteria are stupid, even though they kill us. Evolution is stupid, even though we are a product of evolution. Only human beings are special in this entire universe.

And yet, we still eat, shit, fuck, and get sick and die just like all other living organisms.

The human-centric view of intelligence is fundamentally wrong. The belief that only human beings are intelligent, that consciousness is intelligence—has caused blindness in all those seeking a theory of intelligence.

Patternism will correct the error caused by the human-centric view by confronting the phenomenon of intelligence from two directions:

1. **Expand Intelligence** – Intelligence will be redefined to include all biological phenomena, extending beyond human beings and the human brain. Generations of viruses will be recognized as intelligent, populations of bacteria as intelligent, the immune system as intelligent, and evolution itself as an intelligent process.

2. **Explain All Biological Activities and Interactions** – A *complete theory of intelligence* must explain all regularly observed activities and interactions carried out by biological organisms, including the organisms themselves.

Past understandings of biological phenomena were limited even before excluding human activities and interactions. Too much focus was placed on population genetics, genes, and biomolecular structures related to DNA. This narrow focus neglected a vast range of other activities and interactions that organisms engage in.

V. A Viable Set

The mechanisms of camouflage; the functions of eyes, ears, echolocation, and whiskers; various forms of communication; multicellularity, reproduction, repair, and eating—all must be included. All must be broken down into a coherent set of working principles.

To be called *complete*, a theory must explain all phenomena within its domain. Even within the traditional human-centric approach, the failure to include all activities and interactions persists. Many human phenomena have been, and continue to be, overlooked by those who hold the human-centric view of intelligence. Too much emphasis is placed on neurons, the human brain, consciousness, science, math, and language, while other crucial human activities and interactions are ignored.

Rhyming in poetry, music, and dance. Man-made sensors such as radar, infrared, and microscopes. Blueprints. Lab testing on mice and other organisms. The physical training of athletes by coaches and equipment. Martial arts and the teaching of techniques. Measurement and the use of measuring tools. The phenomenon of simply counting numbers. Propaganda and warfare. Cryptography. Traffic lights. A car signaling with a blinker to indicate a left or right turn. Fishing and the use of plastic lures or metal spoons.

Are these activities not carried out by human beings? Do they not take place in the same universe? Can we not break down all these human phenomena into a set of working principles?

Why, then, is this not done?

The human-centric view of intelligence is a failure. Even if we were to accept the human-centric view of intelligence, why are no attempts made to collect all human phenomena and search for a set of principles within?

It is a massive error. The human-centric view does not even recognize the flaws and inconsistencies within its own framework.

As Patternists, we will correct this error. We will combine all biological and human phenomena into one single collection of phenomena we call *intelligence*. We will find and describe a common set of working principles to break down every phenomenon in our collection.

What if the answer were "*no*?" What would it mean if there were no patterns to be found within the set of all biological phenomena? What does it mean for there not to exist a common set of principles that governs all biological phenomena?

It would mean that every single living organism operates uniquely, with no overlap. It would mean that there is no point in trying to understand living organisms or in pursuing a theory of intelligence or artificial intelligence, as the phenomenon of intelligence would be bound to the biological substrate and be ultimately indescribable.

And yet, we have already found such a set of patterns and used them as principles for a theory. Variation, heritability, differential survival, and reproductive success, we observed these regularities across all living organisms and combined them into the principles that form the foundation of evolution and the theory of natural selection.

However, the theory of evolution—and its modern equivalent—remains incomplete.

While evolution has identified major regularities and successfully explained many biological phenomena, it still lacks inclusiveness. Many activities and interactions in living organisms, as well as the mechanisms by which they operate, have been excluded. Human beings and human activities have been excluded. Ironically, the theory's success in explaining major biological phenomena has led to its stagnation. Scientists and philosophers assumed that evolution was complete—that all biological phenomena had been fully explained.

But they were not.

There is still much work to be done.

Patternism will finish the work and explain all biological phenomena. Patternism will complete the theory of evolution and will replace it.

Through Patternism, we have a method to do so. We have collected our set of phenomena. By applying the *Universal Method* to the collection of all biological phenomena, the regularities of

V. A Viable Set

Representation, *Recognition*, *Reproduction*, and *Randomization* have been identified as a viable set of principles.

A Viable Set of Principles for a *Complete Theory of Intelligence*:

1. **Representation** – Using one set of entities to represent another.

2. **Recognition** – Detecting, comparing, and storing features, leading to the grouping and separation of entities in a consistent manner.

3. **Reproduction** – The propagation of patterns.

4. **Randomization** – Generation of variation as a means to search for patterns.

The goal of this *viable set* is to break all biological phenomena into a working set of principles. That given any biological phenomenon, we can analyze it and observe aspects of *Representation*, *Recognition*, *Reproduction*, or *Randomization* at play, in the same manner that all other successful scientific theories have done with the phenomena within their respective fields.

We are not reducing for the sake of reducing. We are reducing to solve problems and to build and create. The principles in this *viable set* can be applied to any problem we encounter. Instead of proceeding blindly, we now have general principles—universal tools—to dissect a problem or integrate into solutions. When building and engineering, these principles can be universally incorporated into any system, regardless of medium or domain.

This set—a *viable set*—is not the end goal. The importance lies in the attempt to find such a set. Should additional principles be discovered, they will be added. If a better set of principles capable of accounting for more biological phenomena is found, this specific set will be replaced. Patternism is the belief that such a universal set exists. The set presented here is merely viable.

The set presented here represents only secondary principles. We can stop at *Representation*, *Recognition*, *Reproduction*, and *Randomization* and be satisfied that we have identified an effective set of principles covering all biological and intelligent phenomena. But we can go further.

We can ask: "Is there something more fundamental?" Can *Representation*, *Recognition*, *Reproduction*, and *Randomization* be further reduced?

For a Patternist, the answer is *"yes"*—there is a more fundamental, first principle: *Patterns*.

It will be shown that *Representation* can be reduced to patterns. *Recognition* can be reduced to patterns. *Reproduction* can be reduced to patterns. And *Randomization* itself is either used to discover patterns or to counter other pattern-seeking agents.

Patternism is the belief that what ultimately lies at the core of all biological and intelligent phenomena are patterns and the *Paradox of the Pattern*. Only by fully exploring and understanding what a pattern is—both philosophically and practically—can a *complete theory of intelligence* emerge.

Principle 1. Representation

Representation is the act of using one set of entities to represent another set of entities.

A map represents an area of land. A globe represents the Earth. A barcode represents an item in machine-readable form. A person's name represents their identity, and a passport or ID card is a further representation of that individual.

The Mona Lisa represents a woman, a statue of Lincoln represents Lincoln, and a plastic skeleton in a classroom represents the human skeletal anatomy. In representation, a flexible medium—whether oil on canvas, marble, or plastic—is modified to share key features with a target object, persuading observers to categorize it under the same name.

Representation is a fundamental process utilized by all living organisms.

Territorial markings represent an area claimed by a predator. Genes represent protein structures. The bodies of *Phasmatodea* insects mimic fallen leaves and twigs. An octopus camouflages itself by using its body to represent the surrounding environment. Modern military camouflage operates under similar principles.

Laboratory animals represent humans in testing new drugs, as their biological response mirrors those of people. The homunculus in the human brain is a collection of neurons that represent various parts of the human body. An elected official represents constituents, and currency represents resources—whether gold, oil, or quantities of rice.

In chemistry, structural formulas graphically represent molecular structures and bonds. The theory of evolution is a representation of the natural processes by which organisms evolve and develop in complexity. The abstract of a scientific paper is a representation of the full text. Gödel's proof of incompleteness uses representation by expressing mathematical statements as numbers. Any continuous function can be represented as a summation of sines and cosines with different frequencies and amplitudes. The same applies to linear regression or any other regression model.

All theories—whether scientific, political, economic, or religious—are representations conveyed through language. Artwork, actors, video games, textbooks, and novels are all representations. Allegory and metaphor, such as Aesop's fables, represent various aspects of human nature and their resulting interactions. History represents the past, and our memories represent our personal experiences. The *King James Bible* is a representation of the Latin Bible. All translations are representations. When an image is stored on a computer, it is represented as a series of binary code.

The applications of representations are diverse. Representation can store information, as in hard drives or textbooks. It can capture an image, as in a painting, photograph, or statue. It enables communication, as in emails, spoken language, or Morse code. It aids reproduction, as with DNA and blueprints. And it can define features for an entity, as in measurement.

Representations are used to run simulations and make predictions by manipulating variables, as seen in physics and supercomputer modeling. They play a key role in training, such as when astronauts practice in a free-falling airplane that mimics zero gravity. Similarly, pilots use flight simulators to prepare for real-life scenarios. A safe practice environment is a representation of the actual environment.

Representation can serve as a survival tool, as with camouflage that deceives a predator's recognition system. It is used in hunting, such as when a fisherman uses a plastic worm, or a hunter places a decoy of a mallard duck or turkey. It is used in performing mathematical proofs and making theoretical predictions. Lastly, representation can serve as a means of entertainment, eliciting excitement, fear, or emotional responses in a safe environment—such as in movies, plays, or even pornography.

The list of representations is extensive, as are their uses. All these representations exist within the same universe. By appealing to universality, we can conclude that there is an underlying structure that governs all representations.

V. A Viable Set: Representation

What is that structure? What are the necessary components that allow representations to occur? What are the principles that govern all representations?

We apply the *Universal Method* to find our answers. We already have a vast collection of phenomena related to representations—everything we've listed and more. By searching for regularities within this collection, we can identify the essential components that give rise to the phenomenon of representation.

Components of Representations:

1. **Representative Form** – This refers to a set of entities that serve as the medium of representation. Examples include paper and ink, neurons, drawings, photographs, symbols, words, numbers, equations, verbal language, DNA, RNA, TV screens, driver's licenses, maps, statues, paper ballots, figures in tables or graphs, and the body color of an organism. Ideally, this form uses a medium that is easy to manipulate or has distinct features that make it advantageous. On its own, the *representative form* holds little value without a translation system to interpret it.

2. **Translation Machinery** – This system converts the *representative form* into something functional. For human-related representations, this includes the entire human being—not just the brain and senses but also actions involving tools, machines, and social institutions. In the case of DNA, the *translation machinery* includes RNA, enzymes, and cellular structures that convert genes into proteins. For camouflage, the predator's recognition system acts as the translation mechanism, but it is deceived by the prey's camouflage, preventing detection and response. All translation systems are recognition-based, converting features of the *representative form* into an *original form*.

3. **Original Form** – This is the final form of the representation, ready for direct use. It can be a protein, a physical trait, or a fully developed organism after gene translation. It can be a built aircraft from a blueprint, an emotional response to a film, or avoidance of predators. It could include the skills acquired through training, enabling performance in high-stakes situations. Generally, the *original form* is considered to exist in *"reality"* or the *original space*, where constraints are greater, and modifications require more resources and energy.

This framework is completely general, with no inherent limitations. The three components do not need to exist within the same entity. Translation can flow in multiple directions, from the *representative form* to the *original form* and vice versa. The translation does not have to be exact or one-to-one—imperfections, fuzziness, and variations can exist and may even be beneficial.

There is no rule prohibiting a representation from representing itself. A representation may contain instructions for its own reproduction, and representations can be nested within other representations, creating layers that can translate back and forth. The number of layers is unlimited, and in some cases, the *representative form* and *original form* may become interchangeable.

Feedback loops can also arise where the *original form* influences the *representative form*, which then feeds back into the *original form*, creating a continuous cycle. When the *translation machinery* is introduced into this loop, feedback can come from multiple directions. This feedback may involve factors such as energy efficiency, survival, prediction accuracy, correct classification, or effective categorization. Such interconnected loops enable the entire system to optimize, correct errors, manage complexity, and evolve over time.

Although we often treat *representative forms* as existing in a **"*representative space*"** for manipulation and the *original form* as existing in an **"*original space*,"** there is no complete isolation. All three components of representation exist within the same physical reality and are thus subject to the forces of natural selection.

V. A Viable Set: Representation

The ability to layer representations on top of one another has profound implications for the development of a theory of intelligence. It can be demonstrated that all representations are fundamentally the same and can be translated from one form to another, provided strict consistency is maintained and the rules governing the transition between representations are clearly defined—that patterns are followed. This principle underlies the *Universal Turing Machine*, computer simulation, data storage, and the foundation of mathematics.

The result is that intelligence is no longer bound by the biological substrate. It becomes medium independent—describable and translatable into language, genetics, neural circuits, and silicon transistors. Intelligence can be broken down into principles that apply universally. This extends to all theories, as they must be stored in textbooks, computers, and within the human brain.

Representation is a key universal tool.

What has been presented here is completely general and can be applied in countless ways, limited only by the creativity of those who seek to understand representations and use them to build and create.

The first question to ask is: "What are we using the representation for?" Is it a means of storage, searching, accounting, aiding reproduction, simulation, testing, or communication? What specific advantage does the representation provide?

The advantage of a representation depends on the properties of the medium used. Assuming there is a functioning *translation machinery*, the usefulness of a representation relies on the inherent features and capabilities of its medium.

For instance, if a representation is intended for storage, the medium should be robust, compact, and be able to preserve its form over time. For communication, we look for a medium that is lightweight and can travel long distances with minimal energy. If only visual sensors are available, the medium should have visual properties; if audio sensors are involved, the medium must produce sound.

When aiding reproduction, the medium should have a natural mechanism for replication. A representation used for training must replicate the properties of the real event while being safer—

particularly if failure in the live event could result in death. For testing purposes, such as drug trials on lab animals or crash tests with dummies, the test subjects must share key properties with the target subject. For simulations or predictions, the medium should be easy to manipulate, allowing outcomes to be calculated, stored, and compared efficiently. A fishing lure, for example, should mimic the color, shape, and movement of the target fish's prey.

Once the suitable medium with the desired properties is identified, the next step is to set up the *translation machinery* to convert the *original form*—whatever it is we aim to store, reproduce, communicate, or test—into the *representation form*. We then carry out the necessary processes of storage, communication, or simulation until the desired outcome is achieved, at which point we can convert back to the *original form*.

Finally, we verify the accuracy of the representation system by testing the *original form*. We check whether storage was successful and if the information can be recalled when needed, whether reproduction yields functioning copies, whether communication is clear, and whether theoretical predictions align with reality.

There is a ladder that must be climbed.

What separates Patternism from all other theories of intelligence is the accounting of all biological phenomena. With a clear understanding of representations, we can examine genetics as a system of representation.

All living organisms make use of DNA representation. Every bacterium, unicellular and multicellular organism, and even viruses rely on DNA to produce the protein structures necessary for survival.

The question to ask is: "*Why?*"

Why use DNA sequences—combinations of A, T, C, and G—to represent proteins? Why does a codon, a set of three nucleotides, code for a specific amino acid? Why is there a structured table showing which codon corresponds to which amino acid? Why does this system involve two layers of representation, where DNA must first be translated into RNA before the proteins are synthesized? How is it that

V. A Viable Set: Representation

an entire organism can be represented as a long-coded sequence of nucleotides?

What advantages does DNA representation provide, and can these advantages be generalized?

A theory of intelligence that cannot account for the mechanism enabling DNA representation—cannot discuss or even recognize the advantages of DNA representation—is worthless. For a Patternist, representation is a powerful tool that living organisms have used for over a billion years to survive, evolve, and ultimately produce human beings and the human brain.

DNA representation offers three key advantages: ease of storage, reproduction, and manipulation.

1. **Ease of Storage** – DNA provides efficient storage for protein structures and, ultimately, the organism's physical form.

A cell has physical and energetic limits. It cannot maintain all the proteins it will need throughout its life cycle. Some proteins are required only at certain stages or under specific conditions, while others are unnecessary because they are produced by different cell types elsewhere. Certain proteins, if activated at the wrong time, could even interfere with others, potentially causing harm.

Maintaining all proteins simultaneously would be inefficient. Space is limited, and protein structures naturally decay over time. A more efficient approach is for the cell to produce only the necessary proteins while storing the rest in DNA form. DNA serves as a library; as long as there is a translation mechanism to convert DNA sequences into the corresponding proteins, the cell conserves resources and energy by producing proteins only as needed. This efficiency is even more pronounced in multicellular organisms, where each cell type specializes in producing specific proteins, even though all cells contain the complete genetic library.

2. **Facilitate Reproduction** – DNA representation makes reproduction far more feasible.

Reproduction without some form of representation is exceedingly difficult. The only known example of self-contained reproduction without representation is self-assembling prions, which are microscopically small and simple. But how could something as large and complex as an elephant, whale, or giraffe reproduce?

Reproduction becomes more manageable when it begins in a *representative form* and is then translated into the *original form*. In DNA, the genetic code acts as this representative form—it is simpler, easier to manipulate, and naturally prone to replication compared to the complete organism. DNA replicates first, and then the *translation machinery* uses it to produce protein structures, which ultimately develop into the entire organism.

3. **Enable Manipulation** – An organism can be altered by changing its DNA structure rather than directly modifying its proteins or body.

By altering DNA, we can effectively change the organism, with the added benefit that these changes are stored and passed to future generations during reproduction. If an organism were modified without changing its DNA, it would require direct physical intervention, such as surgery or other manual manipulation. This approach would be labor-intensive and be impermanent, as the changes would not carry over to the next generation.

Through DNA representation, genetic engineering has become possible. In nature, DNA changes through random mutation, meiosis, and sexual reproduction. Human beings, recognizing the pattern in which offspring share traits with their parents, have long used DNA representation indirectly through selective breeding to domesticate animals and cultivate crops.

In the modern era, advances in genetics have enabled humans to directly manipulate DNA. Tools to sequence, cut, edit, and insert genes

V. A Viable Set: Representation

allow us to modify living organisms. The process begins by selecting a desirable trait in an organism, identifying the specific genes responsible, and then copying and inserting those genes into a new organism's DNA. The new organism will then display the same desirable traits. If the genes are inserted into germline cells, these changes are passed on to future generations.

This manipulation of DNA representation has enabled the creation of bacteria that produce insulin, the transfer of jellyfish bioluminescence to mice, and the development of insect- and herbicide-resistant crops.

The advantages of DNA representation are all rooted in patterns. DNA itself is a pattern, with each specific gene consistently encoding the same protein structure. The translation machinery that converts DNA into proteins relies on predictable, regular interactions between codons and amino acids. The genetic code is a structured table of these patterns. It is this internal consistency and reliance on patterns that make DNA such an effective system for storing, replicating, and modifying life.

This reliance on patterns is not unique to DNA. It applies to all forms of representation—whether in genetic code, language, maps, or computer systems—representation hinges on the existence of consistent, repeatable patterns that can be translated into meaningful, *original forms*.

Representation can be defined as the act of using one set of patterns to represent another set of patterns.

The reliance of representation on patterns is grounded in practicality. We seek effective representations that are reliable, error-free, and yield useful outcomes. This requirement imposes the need for both internal and external patterns in any system of representation.

Roles Patterns play in Representation:

1. **Internal Patterns** – The representation system must maintain internal consistency. The *translation machinery* that converts the *representative form* into the *original form* must follow a consistent process. For instance, in Morse code, the relationship between letters and their corresponding dots and dashes must remain stable for communication to function correctly. If the mapping between letters and their codes changes or becomes inconsistent—if a *"dot"* no longer represents the letter *"E"*—communication breaks down. The Morse code table itself is a structured pattern that both the sender and receiver must follow.

   This principle also applies to genetic code. In the codon table, each codon must consistently correspond to the same amino acid. If codon assignments changed unpredictably, the proteins synthesized would differ drastically, disrupting biological processes. It is not the specific symbols or sequences that matter but their consistency. For example, a *"dash"* could represent *"E,"* or *"AGA"* could code for *serine*, as long as the new rules are strictly followed.

2. **External Patterns** – Representation must capture an external pattern to be useful. Usefulness depends on reliability, repeatability, and effectiveness—all of which require a pattern to be observed or leveraged.

   When using a representation to store information about an object, the object must be consistent enough to allow accurate storage—whether through filming, recording, or painting. If the object changes too rapidly, accurate storage becomes impossible. Similarly, camouflage works only if the environment remains stable long enough for the organism to adjust its coloration. If the environment changes before the organism can adapt, the

V. A Viable Set: Representation

camouflage fails, wasting energy and increasing the risk of detection.

The same principle applies to maps, which serve as effective representations of land only when the terrain remains consistent enough to align with the map for navigation. This principle extends to naming, labeling, language, training, and all other forms of representation. To ensure reliable communication and functional outcomes, external patterns must be stable and consistent enough for the representation to retain its relevance.

A representation must **synchronize** with what it seeks to represent. A representation is accurate as long as the patterns match. More specifically, a representation is accurate when the features of the *original form* and its *representation form* match. To verify this alignment, we apply the tools of pattern recognition.

We check for repetition—by repeatedly translating the same *original form* into its representation and back, we can assess the consistency of the translation process. We check for categorization, ensuring that the separation and grouping of entities in the *original space* match those in *representation space*, with the *Differentiation of Outcomes* being a key factor. Finally, we check for prediction accuracy, confirming that predictions made in *representation space* translate into predictions in *original space*, which can then be tested and validated.

A feature is anything that can be used to separate and group entities in a consistent manner. When checking for feature-matching between the *original form* and its representation, we're not comparing physical attributes like red-to-red or blue-to-blue—we are not comparing within the same system. Instead, we are checking whether the grouping and separation of sets and subsets remain consistent across systems. The question becomes whether we have a complete translation mechanism that accurately maps sets in the *original space* to corresponding sets in *representation space*.

For instance, consider a simple case with two entities in each space: a *representation space* of "*switch on*" and "*switch off*," and an *original space* of "*lightbulb on*" and "*lightbulb off*." The task is to determine the mapping of what "*switch on*" and "*switch off*" correspond to. This is the job of the *translation machinery*. Once the mappings are established, they must be consistently followed. If the machinery assigns "*switch on*" to "*lightbulb on*," then every time we see "*switch on*," we should expect to see "*lightbulb on*" in the *original space*. If we find "*lightbulb off*" instead, we assume an error has occurred.

The result of this process is *synchrony*. Synchrony allows us to treat layers of representation as a **black box**, providing a powerful tool for understanding complex systems. For example, we say that DNA affects an organism's traits, but the gene must pass through multiple layers of processes: translation to RNA, protein synthesis, and ultimately influencing a trait through complex molecular interactions. While the link from gene to trait is indirect, as long as each layer functions consistently, we can treat the system as a black box and practically conclude that the gene directly affects the trait.

This synchrony also applies to the homunculus in the human brain. Sensory signals travel from the fingertips, through nerve cells, up the spinal cord, and to the brain for processing. We can largely ignore the intervening nerve cells as long as they function correctly—when one turns "*on*," the next in line turns "*on*," and when one is "*off*," the next remains "*off*." When this synchrony breaks down, as in nerve damage, we experience numbness or paralysis, or, if cells misfire, random pain or seizures.

The same principle applies to Morse code. There are 26 distinct codes for letters and 10 for numerals. Whether conveyed visually with flashlights, auditorily with tones, or by taps on a wall, accurate communication is achieved as long as the medium synchronizes with the code. But how is this possible? What allows Morse code to function independently of the medium?

It is the separation and grouping of entities in a consistent manner that is key to representation. When we examine language as a system of representation—where words serve as labels for groups of entities

V. A Viable Set: Representation

that a society interacts with—we see that the specific labels, symbols, or vocal sounds are less important than maintaining clear differentiation between them. For instance, if a dictionary is a list of patterns that a society finds useful, it doesn't matter whether the dictionary is written in English, Russian, or Chinese. Apart from specific cultural nuances, we expect common groupings of words across societies, such as those for *"mother," "father," "family," "food,"* and *"water."*

Moreover, we expect relationships between words to exhibit patterns. For example, words such as *"food"* and *"water"* will appear in the same sentence more often than *"food"* and *"cloud."* The conceptual distance between *"food"* and *"water"* is closer than that between *"food"* and *"cloud."* This relational mapping applies to all words in a given language, and we expect these relational distances to align across different languages. It is the relationships between words that matter, not the exact labels used. Consistency is key. If the same word is used to describe different entities, confusion arises, necessitating further clarification.

Since different languages aim to describe the same external patterns, we expect there to be common groupings of words across languages. Given this commonality, we expect there to be an effective *translation machinery* that can translate from one language to another—map representation system to representation system—with high accuracy. Thus, we have the phenomenon of translation dictionaries that connect English to Russian, Russian to Chinese, and so on. Even historical artifacts like the *Rosetta Stone*, which translated Ancient Egyptian into Ancient Greek, show that translation between representation systems is possible due to these shared underlying structures.

This leads us to ask: "What allows for the independence of **representative tokens**?" If we strip away the specific symbols, sounds, and labels tied to a particular language, and do the same across multiple languages, what remains? What is the underlying structure that all language shares? What is it about the structure that allows the medium—whether visual, auditory, or otherwise—to become secondary, enabling the independence of representative tokens?

It is the *separation and grouping of entities in a consistent manner* that matters most in representation. This structure—the way entities are grouped and separated, along with the relationships between these groupings and separations—is what allows representative tokens to exist independently of the medium. **Universal Representation.** If a representation system groups and separate entities in a way that matches how reality groups and separate entities, then the representation system will be accurate and effective.

Ideally, when creating a representation, we would have direct access to the complete set of all possible ways entities can be categorized within a given *original space*. With such access, we could catalog all groupings and separations and map the *original space* to a *representation space* one-to-one, similar to how an explorer creates a map by directly traversing land and recording observations at each location.

In practice, however, we rarely have access to the full set of categories. Even if we did, the sheer number of potential categorizations is often too vast to record. We lack the necessary sensors and ability to observe every location, and we do not have enough neurons, symbols, paper, or silicon transistors to store all possible categorizations.

In such cases, we turn to patterns. We predict and then confirm. By identifying the *"**rules**"* that govern how entities are grouped and separated in an *original space*, we can translate these rules into a *representation space*. This allows us to generate new representative groupings—new representative entities—that can be used to search for corresponding groupings in reality. In essence, we apply the *Universal Method* to establish principles for a theory.

Usually, we uncover these rules by sampling from the *original space* and employing pattern recognition—applying *Repetition*, *Prediction*, *Natural Selection*, and *Categorization*, along with direct experimentation, to identify the patterns that define the grouping and separation rules for the new representation.

V. A Viable Set: Representation

The only requirement for these rules is that they be consistent and regular, following a calculable and reproducible pattern. While perfection is not necessary, the rules must achieve a high degree of approximation. If the rules governing groupings and separations are inconsistent or incalculable, no meaningful groupings can be generated. The results would be too random to serve as useful representations, with too many potential outcomes and branches to manage.

Ultimately, we can ask: "What kinds of rules govern the universe?" Some rules may be inherently inconsistent, defying patterns altogether and thus remaining unrepresentable, incalculable, and beyond capture by language, mathematics, or the human brain. Other rules may be semi-consistent, following patterns only within specific domains. In such cases, representation is possible, but the rules lead to branching outcomes, resulting in inconsistent groupings. Finally, some rules are highly consistent and regular, adhering to strict patterns that can be represented with symbols, manipulated, and used for computation with incredible predictive accuracy—giving rise to mathematics and physics.

These consistent and regular rules, once identified, serve as the foundation for creating representations that are both meaningful and reliable. The more consistent and regular the rules, the more effectively we can use them to categorize and separate entities in a way that reflects reality. This leads us to one of the most direct and practical applications of such rules: measurement.

Measurement is the representational definition of features for a given real-world entity. It is a natural extension of the principles discussed, involving the use of consistent rules to define and represent features. By labeling an entity, measurement facilitates categorization, enabling us to group, separate, and manipulate its representation using symbols and figures—without needing to interact with the real-world entity after the measurement is taken. This allows us to handle objects' properties more efficiently, making comparison, manipulation, and analysis easier than working directly with the physical objects themselves.

Measurement begins with the establishment of a fundamental unit. For measurement to occur, a well-defined fundamental unit must be part of the *translation machinery* that connects real-world quantities to those in a *representational space*. Typically, a highly consistent real-world object is chosen as this unit, such as the original meter stick or pound.

However, it is the consistency of the fundamental unit that is critical, not its physical form. Consistency acts as both the *rule* and the *ruler*, with the physical form simply ensuring the unit remains unchanging. In modern times, for greater precision, fundamental units have been redefined using entities so consistent that they are considered constants—like the speed of light or the charge of an electron. When combined with the phenomenon of counting, measurement gives rise to diverse units and systems, along with methods for converting between them.

How are new entities, groupings, and separations derived using the defined rules within a given *representation space*? If we momentarily set aside reality and focus solely on interactions within the *representation space*—concentrating on how rules generate new representational entities—we enter the realm of mathematics.

In mathematics, the primary concern is maintaining a high level of consistency when applying rules to derive new entities. Having multiple answers or outcomes would lead to confusion and so we want consistency. We want a single answer, a single prediction, a single outcome, a single derived entity. Once an entity is categorized—grouped and separated from others—we expect that categorization to remain consistent every time.

The defined rules—the set of **axioms**—govern how entities in a *representation space* are manipulated and how they interact with one another. If mathematics is treated as the manipulation of symbols according to a set of axioms, certain questions arise: "Are these symbols equivalent to others, or do they represent separate entities?" Given a group of symbols, can we manipulate them according to the axioms to match another group of symbols on the opposite side of an equation? For any valid statement to which axioms can be applied, is

V. A Viable Set: Representation

there a consistent, final answer as to whether they belong to the same group or a different group? Will this final answer remain consistent every time, regardless of how the axioms are applied?

We can also question the axioms themselves. What is the complete set of axioms for a system, and are they well-defined? Does the order in which the axioms are applied affect the outcome? Can a better set of axioms be found? Is it possible to break an axiom down into more fundamental principles, creating a more foundational representation?

Then there are questions about inherent inconsistency, particularly with self-referential entities. Is the self-reference of an entity the same as the entity itself? Does an entity and its self-reference belong to the same group, or does it stand apart? Is a set that contains itself truly a new set, or just a continuation of the old one? Are self-referential statements valid at all?

These questions extend to the nature of representation itself. Is a representation identical to what it represents, or does it constitute something new? Should self-referencing and representations be permitted in mathematical proofs?

By addressing these questions, we touch the very roots from which all other branches of mathematics—set theory, category theory, number theory—emerge. What we've shown is how patterns, features, and the grouping and separation of entities underpin mathematics. Through Patternism, we demonstrate that mathematics is fundamentally the study of patterns.

Now we connect representation to reality. When we derive a new entity in the representation space, we are essentially making a prediction about the existence of corresponding entities in reality, which we can then search for. We predict where these entities should be located and what their distinguishing features might be. If the rules governing our representation accurately reflect patterns in reality, we should be able to find those entities, along with the predicted features as translated from representation.

This interaction—identifying patterns in reality, using those patterns as axioms or principles in a representation, and generating predictions about real-world entities to search for—forms the foundation of physics and all theoretical science.

The question we now pose is: "What are the limitations of what can be represented?"

This question can take many forms. We ask: "What can be consistently categorized with measurable results?" What can be represented by neurons, words in a book, or as genes?

What kinds of functions are computable in mathematics? What kinds of functions can be represented by sines and cosines? What kinds of data are suitable for regression analysis? Can everything in the universe be reduced to simple conditional statements like "If A, then B," or are there irreducible cases of "If A, then *sometimes* B" or "B is completely random?"

What are the limits of sensors that can be built? Are there entities beyond our capacity to detect? Are there aspects of reality that we will never be able to "*sense*?"

What are the limits of representation?

Representation is the act of using one set of patterns to represent another set of patterns.

What can be represented with any degree of accuracy are patterns. It does not matter what the entity is—a force, an event, a movement, a measurement, a relationship between entities, or physical objects. As long as the entity exhibits regularity, it can be represented or at least approximated with accuracy.

To calculate is to use patterns. Whether it's measuring, simulating, predicting, counting, manipulating symbols according to axioms, following rules, applying principles, or generating text or images, these actions all result from applying patterns. The medium—be it numbers, formulas, words, sentences, paragraphs, or even beads on a string—does not matter. If a pattern exists, it can be used to calculate. Mathematics, language, physics, and machine learning algorithms—such as those in Large Language Models or image recognition—and tools like the abacus, all utilize patterns to perform calculations. Ultimately, the phenomenon of calculation itself arises from patterns.

V. A Viable Set: Representation

A representation exists in the same universe as what it seeks to represent. While a representation may be treated as if it operates within its own isolated space, it remains part of the broader universe and is therefore subject to the same evolutionary processes of variation and adaptation.

This evolutionary process drives the development of increasingly refined and accurate representations. Different representational systems emerge with varying levels of consistency, but only the most effective, reproducible, and reliable ones persist. Representations that accurately capture patterns survive, while those that fail go extinct. Since patterns themselves change over time, representations must evolve to keep pace with shifting external realities.

Consequently, representations require continual updates. Organisms adapt to changing environments, and their genes shift to reflect new conditions. Maps must be revised as new cities and roads emerge. Scientific theories need ongoing testing and re-evaluation, particularly when new phenomena are observed.

And there is always *The Divide*—the gap between a pattern and its recognition, between a pattern and its representation. Do not confuse the representation with what it seeks to represent. A painting of a flower is not the flower. A ballot is not the voter. A deed to a plot of land is not the land. A sequence of DNA is not the organism. Money is merely colored ink on paper. A documentary is not the actual event. What is written in a nation's constitution may not reflect what is truly upheld and applied. When we mistake the representation for reality, we run into trouble. When we fail to consider the entire representation system in its totality—along with all the components that allow it to function—we run into trouble. We must remain aware of what constitutes the *representation form*, the *original form*, and the link between them—the complete *translation machinery* and all its aspects.

Intelligence is the ability to find and utilize patterns. If an agent is intelligent, it will naturally seek out patterns to exploit. This is the absolute prediction of Patternism. Since representations are full of patterns, they offer numerous opportunities for exploitation—particularly because targeting the representational form is often easier and more energy-efficient than attacking the reality it represents.

A predatory intelligent agent will find any weaknesses in the representational form of any given representation system to further its own ends. The phenomena of viruses, genetic modification in the biotech industry, and the counterfeiting of money. Videos and images are manipulated to shape narratives, and biased data are presented to sway opinions. The manipulation of language becomes propaganda and disinformation, whether propagated by governments, media, or individuals. All are instances where representations are exploited by intelligent agents.

If this exploitation occurs often enough, it becomes a regularity—a pattern. If the representation system is intelligent or connected to an allied intelligent agent, it will eventually evolve countermeasures to reinforce its representational integrity.

The adaptive immune system evolved to combat viruses that hijack the cell's translation machinery. Government agencies are established to monitor counterfeiting. Checksums and scans are implemented. Bugs are patched in new software versions. The field of computer security exists to counter hacking and phishing. Auditing ensures that records align with reality—that physical items in a warehouse match what is stored in an inventory database. The physical counting of cash at day's end and comparing it to the list of items sold, auditing company assets, departmental budgets, and financial reports—all serve to maintain representational integrity.

While these safeguards increase the energy costs of maintaining a representation and reduce efficiency, they are essential. A representation system without safeguards may function for a time, but in the presence of intelligent, predatory agents—especially human beings—it will inevitably be exploited. Ironically, if the predatory agent relies too heavily on exploiting a particular representation system for survival, it risks collapsing along with the representation it exploits—unless it adapts to become less destructive.

Things will break. If a representation is overexploited to the point of losing its connection to reality, the entire ecosystem can be put at risk. Errors will compound, trust will erode, and individuals will learn to disregard the representation entirely. Lies will become unsustainable, economies will collapse, predictions will fail, and resources devoted to maintaining the representation will be wasted. Systems will break down until sanity is restored—until patterns are restored, and the representation once again aligns with the reality it seeks to represent.

We apply everything discussed about representation to the human mind. In absolute physical terms, everything comprehended by human beings must be filtered through the brain and, thus, represented as a network of neurons. Every object we see, every concept we imagine, and everything we perceive exists in the form of interconnected neural representations.

Ultimately, all that we have access to are representations.

We create representations of the world around us, building an entire universe for ourselves. We hold beliefs about our reality and how it functions. We form mental maps of our local geography, such as the location of our home and how to reach it, or the roads that take us to work or school. We hold beliefs about which direction is up, so we don't fall. We create mental simulations of where a golf ball, an arrow, or a frisbee will land. We have moral and philosophical beliefs about how society and the universe should function. This applies to all objects and concepts we recognize, categorize, and name, including other human beings.

Our internal representations don't have to be perfect—only good enough to survive. At times, our individual representation of the world strays too far from reality. We might even choose to ignore reality altogether and live entirely in the universe we've created for ourselves. And if the environment is not overly selective in its elimination, we can get away with it. For a time, we can live within our own illusions—particularly in the modern era, where resource abundance and technology allow a representation to diverge from reality before reality snaps back.

But reality always seems to snap back.

Small divergences can be corrected without much stress. Large divergences cannot. Correcting a large divergence will break a man. It will break society. Yet this is what it means to reason.

Human beings have a unique ability to change reality to match our representations. We can imagine potential realities and work toward making them real. We create visions in our minds—representations of a potential future to which we can dedicate our time and energy. We write about these visions, share them with others, and plan the steps to achieve them. As long as the vision is realistic, we have a chance of accomplishing it. But this requires hard work and an accurate understanding of current reality.

Constant input is essential. Shaping reality to match a vision requires a steady flow of high-quality feedback, allowing us to compare our vision against reality. This process demands a discerning eye, attention to detail, and a willingness to adjust our vision when necessary. We take a step, observe reality, and reevaluate to determine whether our actions have brought us closer to or further from our goal. Then we take another step, observe, and reevaluate—again and again.

If we ignore reality at any point or become too seduced by our vision, we risk becoming ideological. We rationalize instead of reason—often without realizing the difference—and risk falling deeper into madness.

We apply representation to the self. Just as an individual can have a representation of the world around them, we can have a representation of ourselves.

We see ourselves as a pattern. Our body and what is reflected in a mirror. The set of neurons called the cortical homunculus inside our brain represents the various parts of our body. Our name, background, identity, and family. What belongs to us and what we own as individuals. What we can control and what we can reach out to stop. All comes together to compose the representation we call the *"self."*

With this representation of ourselves, we can predict our own behavior. We can run simulations of ourselves within our minds, interacting with our current representation of the world, and evaluate various outcomes to help us decide our actions in reality.

V. A Viable Set: Representation

We see ourselves as an evolving pattern. Over time, we change, and so our representation of ourselves must also change. It must evolve. We create a representation of our future selves—a self-potential of who we want to be. What we might look like if we exercise and change our diet, improve our skills and abilities, or develop our character to become a person of courage. We imagine achieving great things, finding purpose, choosing a career, deciding whom we will marry, how many children we'll have, where we'll live, and what kind of house we'll call home—a vision of who and what we want to become.

Just as we can shape reality to fit our representations, we can modify ourselves and work towards a vision of our self-potential. But this, too, requires hard work and constant feedback from our current reality.

Every day when we wake up, we must ask ourselves, "Am I closer to, or further from, who I want to be?" Time is infinite, yet it is not. The life of a single man is incredibly short. Steps must be taken toward the goals we set for ourselves, and each step must be evaluated with brutal honesty. To take no steps at all is itself, a step. And the life of a single man is incredibly short.

If we are closer, we smile, knowing that the pain we wake up with comes from the hard work we did yesterday—that our body and mind are in the midst of a transformation for the better. If we are further, we must face the knowledge that we have been lazy, and it pains us even more. A different kind of pain—a pain that strikes us to the very core of our being. And we feel it: the guilt, haunting us when we are alone in the twilight hours.

For this is what it means to suffer—and the difference between suffering meaningfully and meaninglessly.

Principle 2. Recognition

Recognition is the process of detecting, comparing, and storing patterns.

Since a pattern consists of features, recognition involves an agent identifying, comparing, and storing these features to determine when to trigger an action. The process of recognition is dynamic and ongoing. It requires the agent to continuously monitor its environment, filter relevant features from irrelevant ones, and adapt its responses based on new information or changes in context.

The phenomenon of recognition is extensive. It involves determining whether an approaching entity is a human or an animal. If it is an animal, recognizing whether it is a cat or a dog, whether it appears aggressive or shy. If it is a human, identifying whether the person is a friend or foe, a member of the same tribe, or an outsider. Recognition includes distinguishing a tiger from a housecat or a wolf from a golden retriever, as well as recognizing similarities between them. It also includes identifying specific individuals, such as celebrities, family members, or coworkers.

When a tree is felled in a forest, we separate insects from leaves and further identify different species—whether moths, grasshoppers, and wasps. The entire field of taxonomy is dedicated to recognizing the diverse species of living organisms, cataloging their unique physical and genetic features.

Recognition covers interpreting traffic lights—red for stop, yellow for caution, green for go—and determining how to navigate vehicles accordingly. It includes recognizing the blinking signals of a car indicating a brake, hazard, or an upcoming turn and distinguishing between different types of cars.

Measurement relies on recognition—identifying units like inches, meters, and kilograms. The various rulers and devices used to take these measurements serve as tools of recognition, assigning and associating features to real-world objects.

V. A Viable Set: Recognition

Recognition includes identifying different types of movement: running, walking, and swimming. It involves recognizing techniques performed in a combat sport, including whether they are executed correctly and their impact on the opponent—whether the damage was critical or superficial. It extends to identifying the best among many, such as awarding a gold medal for first place or a silver medal for second.

Recognition includes proficiency in languages, whether English, Vietnamese, Morse code, or technical jargon in fields like machine learning or physics. Each word in a dictionary represents the recognition of a pattern.

Recognition extends to diagnostics and problem-solving, identifying when a current action is ineffective or disastrous if continued. A doctor recognizes a disease or illness based on a patient's symptoms. Similarly, a mechanic identifies car issues by conducting a test drive, listening to engine sounds, or analyzing descriptions provided by the car owner.

Recognition of goals involves determining whether a proposed solution meets predefined criteria. Whether carried out by a detective solving a crime, a vulture seeking carrion, or a spider looking for a mate, goal recognition involves scanning, comparing, and matching features. This process applies across environments—from prairie fields, oceans, and cities to microscopic levels within cells or data on a laptop. Whether searching for food, a mate, car keys, drugs, proof of a mathematical statement, or solutions to equations, the process is fundamentally one of recognizing goals.

Pattern recognition involves identifying when a phenomenon follows an underlying pattern, determining when that pattern has been fully described, and understanding how these patterns can be used to construct a theory. It includes recognizing which theory most accurately reflects reality.

Recognition extends to consciousness, encompassing the ability to determine whether a person is awake, responsive, and alive versus unconscious or deceased. It involves identifying whether a message originates from a human or a machine. Recognition also extends to one's own beliefs: recognizing something as *"true"* enables actions and adjustments to beliefs, while recognizing something as *"false"* prompts skepticism and caution.

And there is self-recognition. Awareness of our body through pain sensors enables us to react to harm. Recognizing the image we see in a mirror, a picture, a video, as well as the name on documents, as representations of ourselves. Identifying objects under our control—whether it's our body and emotions, a vehicle we drive, or an avatar in a video game. Recognizing that external events can be traced back to our actions. Recognizing that we are a pattern that changes and evolves over time.

Recognition is not limited to humans. Other organisms also act as agents of recognition.

A spider senses trapped prey through web vibrations. An octopus adjusts its body to match its surroundings. Animals recognize members of their own species and refrain from fleeing. They recognize potential mates and assess suitability through competitive displays, dances, or songs.

Plants recognize sunlight, unfurling their petals. A Venus flytrap detects when a fly lands using hair-like structures that trigger the trap. Ants distinguish members of their own colony through pheromones and attack intruders.

A neuron receives input signals through its dendrites, processes these in its soma, and sends an output signal through its axon terminal—effectively performing recognition. A virus recognizes a potential host. White blood cells distinguish pathogens from normal cells. Enzymes recognize substrates to catalyze reactions, and protein receptors detect specific molecules, triggering cellular responses.

Recognition extends beyond biology to mechanical sensors and devices.

A thermometer measures temperature to activate an air conditioner. Radar identifies airborne objects, while sonar detects underwater objects. A lock recognizes a key—whether physical or electronic, such as a password, number combination, fingerprint, or retina scan.

V. A Viable Set: Recognition

In machine learning: text recognition, speech recognition, and image recognition are all forms of recognition. Even a simple if-else statement in programming represents a basic unit of recognition. Security systems that set off an alarm when movement is detected, pressure plates, bear traps that close when stepped on, and modern anti-tank and anti-ship mines—all activate when a specific feature is triggered, making them examples of recognition.

To categorize, identify, classify, name, differentiate, perceive, or detect with consistency is to recognize.

All occurrences of recognition can be fully accounted for and generalized in terms of features.

A feature is defined as anything that can be used to consistently separate two entities. Once entities are separated by features, the features can then be used to group the entities under a single name—a category. Thus, a feature is anything that enables both the separation and grouping of entities in a consistent manner.

For example, the defining features of the entities we call sharks include a cartilaginous skeleton, five to seven gill slits, and pectoral fins that are not fused to the head. A distinguishing feature of spiders is their eight legs and shared genetic characteristics. The defining characteristic of a prime number is that its factors consist only of 1 and itself. A key feature of entities called chairs is their suitability for sitting. Several features of a person are listed on their driver's license, such as age, height, eye color, and address. Features are synonymous with descriptions, characteristics, properties, and attributes.

A feature can be anything. While common features include color, size, and shape, there are also time-dependent features, such as movement speed (fast or slow), age (young or old), electromagnetic charge, location, or even a specific genetic sequence. As a result, the ways to define a feature are nearly limitless, with the only requirement being that the feature is consistently detectable.

This openness in defining what can constitute a feature is deliberate. The goal is to develop a general theory of description capable of addressing all questions related to the act of description. What is categorization, naming, and labeling? How does a dictionary work? What does it mean to define a word? How is genetics used to define species? What is a biologist's field journal, and why does it contain images and descriptions that highlight specific morphological traits to correctly identify organisms encountered in the field?

Why do blueprints include arrows pointing to critical measurements and allowable tolerances? How does the act of measuring work, and what does it mean for something to be within tolerance? In manufacturing, what are the consequences if a machined part fails to meet the critical features specified by a blueprint?

How are objects grouped into sets? How do we show similarities and differences? How do entirely new features emerge, get named, and described? What is the purpose of these processes, and how do they function? What common principle unites all these phenomena?

All can be explained in terms of features. All can be fundamentally reduced to the separation of two entities in a consistent manner. Features will serve as our first principle.

We start with a single entity that we want to divide into a collection. First, we separate out one new entity from the original. Next, we repeat the process—separating out a second, a third, a fourth, and so on. Beginning with the separation of two entities, we can logically extend the process to separate multiple entities, building a collection of distinct entities from the original one.

Once we have this collection, we can form groups using the same method of separation. We gather all the distinct entities into a single, original collection. This time, we separate out a grouping of entities, treating this group as a single unit. This process allows us to categorize and label the group under a common category or name.

For example, we can take a forest and break it down into individual entities: a single fallen leaf, a single ant, a single tree. We then gather these individual entities into a collection and separate out groups based on their characteristics. We separate out a group of entities we label as *trees*, another group as *insects*, and yet another as *mammals*. In this way, we transform the forest into a set of categorized groups.

V. A Viable Set: Recognition

This method of separating entities is recursive. Any grouping of entities can be further divided using the same approach, yielding two, three, or more sub-entities. These sub-entities can then be organized into new sub-groups. The process can continue, with sub-entities being further divided into sub-sub-entities, which can themselves be grouped into sub-sub-groups, creating increasingly detailed levels of categorization.

We can take the group of entities classified as trees and further divide them into sub-groups, such as *oak* trees, *birch* trees, or *palm* trees. Similarly, we can break down the group of insects into *ants*, *beetles*, *butterflies*, and more. This process mirrors the phenomenon of taxonomy—the systematic classification of living organisms—where broader categories are divided into more specific ones.

What we have derived is a method for creating sets and subsets of entities based on their similarities and differences. What we have derived is the process through which new features come into existence. What we have derived is the phenomenon of categorization starting from first principles.

The key requirement is that the process of separation and grouping remains consistent and regular—that it follows a pattern. If this process were random, our groupings—our sets and subsets—would also be random, leading to inconsistency and a loss of meaning.

There are no limitations.

We categorized entities in a forest, but why stop there? Why not apply this method to the entire Earth, separating out deserts, mountains, jungles, and oceans, and then identifying the various organisms within each biome?

And why stop at ecology? We can categorize materials, substances, and even technology. We can separate out lithium, distinguish protons, neutrons, and electrons, and identify transistors and resistors. We can classify vehicles, computers, weapons, factories, and satellites, separating them from all other entities that exist on Earth.

But why stop there? Why not apply the same process to the entire universe? We can categorize stars, planets, and galaxies. And why limit ourselves to objects? We can examine temporal and spatial features, distinguishing movement, mass, and velocity. The act of measuring and counting is itself a way of consistently separating entities. Assigning a value to an entity—whether coordinates, velocity, or mass—is equivalent to defining a feature for that entity and labeling it. This leads to the phenomena of precision and tolerance: determining whether an entity meets the criteria to belong in a certain category or is excluded based on whether it passes or fails inspection.

We can extend this approach to mathematics itself, as all mathematics can fundamentally be reduced to the consistent separation and grouping of symbols.

And why stop there?

What we have is every word in every dictionary—a ***Universal Dictionary***, not limited to English, Russian, or Chinese. This dictionary is not confined to biology or any technical field; it contains all languages and codes, from Latin, Egyptian hieroglyphics, and Sumerian, to Morse code and even genetic code.

And why stop there?

We can strip language down to its purest form, treating words as entities to be separated and grouped. We can take a word and break it down into a collection of other words—this is the essence of a dictionary. Every entry in a dictionary is a breakdown of that word into a set of other words. Similarly, we can take several words and group them together to form sentences, combine sentences to form paragraphs, organize paragraphs into chapters, and compile chapters into books. This process would create a ***Universal Library*** containing every book that exists or has ever existed, on any imaginable subject—both fiction and non-fiction. The only requirement is that the way we separate and group words remains consistent, following a discernible pattern.

And is this not true?

In the separation of words within a dictionary, is there not consistency in how words are defined? If there were no consistency, the definitions would be random, rendering the dictionary useless, as it would change each time it is consulted.

V. A Viable Set: Recognition

When grouping words, are there not regularities in how correct sentences are formed? Isn't it adherence to a pattern—subject, verb, and object—that distinguishes a grammatically correct sentence from one that is not? The same applies to paragraphs and chapters. Are there not patterns in how well-structured paragraphs and chapters are formed, rather than being entirely random?

In conversations using language, are there not regularities in how questions are asked, or are they entirely random? Is there a pattern in how we respond to questions based on their nature, or are answers random as well? If we built a database of well-structured questions alongside their expert-judged responses, would we find a pattern, or would the questions and answers be completely random?

The same questions can be asked about books. Are the contents of a well-written book entirely random? If they are not, then they follow a pattern, especially if the book reflects reality.

And is this not true? Every well-crafted book has a subject, a central focus around which its content revolves. Every book can be seen as a pattern, with its chapters representing the pattern's features. Every paragraph within a chapter is a feature of that chapter, and every sentence within a paragraph is a feature of that paragraph. Thus, every book can be seen as a pattern composed of features—otherwise, its content would be entirely random.

The structure of language follows patterns—a *linguistic recognition structure*. If we know that language follows patterns, we can use the tools of pattern recognition to capture that structure within a **linguistic recognition engine**. We categorize by gathering well-written books and articles that reflect the reality that human beings live in, and we use prediction to train and test the accuracy of patterns in the engine. We can further apply natural selection by having multiple linguistic recognition engines compete with one another.

Once we have adequately captured the structure of language within the engine, we can use it as a **Linguistic Algorithm** to calculate and generate text, just as all patterns are used for calculation. If the patterns in our linguistic recognition engine are accurate, the resulting calculations—text generation—will also be accurate, reflecting the same reality on which the engine was trained—the same reality that human beings describe. The engine will group and separate word entities, generating text that mirrors the structure of language and the reality it represents.

The separation and grouping of entities in a consistent manner leads to self-recognition.

We apply the method of the separation of two. We take a single entity and separate it into two groups: "***self***" and "***not-self***." We create one group that contains only the entity itself, and another group that contains all other entities that are not itself. The group labeled "*self*" becomes a category, a name, a representation of the entity—it acts as a pointer that refers only to itself. Through this method of separation of two, we establish a mechanism for self-reference.

Self-contradictions are not allowed. Entities must be themselves: $1 = 1$, $X = X$, an apple is an apple. The entities that a recognition system interacts with must fall into the category of "*self*" and not "*not-self*"; they cannot belong to both categories if the system—the structure in which the entities reside—is to maintain strict consistency. This is the fundamental core of all logic.

In cases where two entities, presumed to be the same, are found to differ and thus cause inconsistency, additional clarification or context is needed to resolve the contradiction. For example, consider homographs in language: the word "*bow*" can refer to both a tool for shooting arrows and the front of a ship.

True or false. The set of all sets. Prime numbers. Gödel's Incompleteness Theorem. "*This statement is a lie*" and other self-referencing paradoxes. The question of a system's *self*-consistency, as well as the act of proving things within the system, all relate to how a recognition system manages the categories of "*self*" and "*not-self*"—how it treats entities that fall into one category, the other, both, or neither.

By connecting a recognition system to an agent within an environment and enabling it to categorize interactions as either "*self*" or "*not-self*", we create a foundation for self-differentiation. When this capacity for self-reference and self-differentiation is combined with an

V. A Viable Set: Recognition

agent that can interact with its environment—equipped with both internal and external sensors to receive feedback from its interactions—we achieve self-recognition. By combining self-recognition with the need for full autonomy, we create a system capable of adapting to its surroundings, learning from its experiences, and making decisions independently. We will be creating a fully intelligent and conscious system.

The separation and grouping of entities in a consistent manner goes far beyond.

It is the same as taking an entity and breaking it down into its component parts or taking component parts and building them into a new entity. It is the same as taking a class of phenomena and dividing it into a set of principles that can be used to craft a theory, which is then applied to discover and explore new phenomena. It is the same as taking images, labeling the objects in each image, and feeding them into a machine learning algorithm for training.

The structure is the same—***Universal Recognition***. It is because of this structure that makes mathematics and language work. It is why physics, science, and measurement work. It is why networks of neurons and Large Language Models work. It is the foundation of how all physical matter operates—where all matter can be broken down and separated into atoms, and atoms can be grouped together to form molecules, compounds, and ultimately the entire material world.

This is the universal structure of patterns and their recognition.

Universal recognition is Universal Representation. Our derivation of *Universal Recognition* using the separation and grouping of entities directly mirrors our derivation of *Universal Representation*—with representational patterns matching how reality organizes entities into recognizable forms. Whether we view it as recognition or representation, we are observing the phenomenon of patterns from different perspectives.

Thus, recognition can occur independent of medium—it transcends the specific form in which it is encoded. This universality allows patterns to be stored across diverse forms, from neural networks and language to genetic material and artificial systems. The storage and replication of these patterns enable their recognition, as stored patterns serve as reference points, allowing systems to identify, compare, and respond to features consistently. This is the essence of recognition itself, where the act of recognizing a pattern is inherently tied to its storage and retrievability across different mediums.

The full consequence of this derivation is absolute. If the human brain is viewed as the ultimate recognition machine—the ultimate categorizer—then any theory of intelligence must reduce recognition and categorization to the activities of networks of neurons and, ultimately, to the interactions of individual neurons.

Under Patternism, the separation of entities aligns with key features of biological neurons. Individual neurons function as *on-off switches*, either firing or not based on input, which can be viewed as a mechanism for differentiating between two states. When neurons connect in networks, they form complex circuits where groups of neurons work together to process and categorize information. In a simple analogy, neurons at different levels in a network may represent subdivisions of finer features or combinations of broader categories.

For this neural structure to operate effectively, it must maintain consistency in the separation of entities. Networks of neurons achieve this by using the tools of pattern recognition. They utilize a prediction mechanism, such as threshold potential, where initial inputs into the network partially activate specific neurons. These semi-activated neurons represent a set of predictive features for an anticipated outcome. When the outcome occurs, its input feeds into the same semi-activated network. If the activated neurons from the outcome match the initially semi-activated ones, they surpass the threshold, fully activating and representing shared features between the predicted and actual outcomes.

Threshold potential serves as a feature-matching mechanism. The more matching features, the more neurons are fully activated, increasing predictive accuracy. If no neurons are fully activated, the initial prediction was incorrect. This threshold mechanism can combine with repetition, where frequently activated neurons and connections are reinforced, making them easier to trigger, while unused connections are eliminated. Threshold potential can also integrate with natural selection algorithms, where connections are randomized, and the network refines itself based on prediction accuracy. Furthermore, an overarching natural selection process occurs as neurons—and the organism they belong to—must optimize energy and resource use, survive, and reproduce. These mechanisms maintain neural network consistency across multiple levels and timeframes.

This is the fundamental ***Neuronal Algorithm*** sought by philosophers and scientists to answer how simple neuron interactions can give rise to categorization, recognition, and ultimately intelligence.

But we can take this further.

Any theory of intelligence must show how all the sought-after products of human intelligence—whether music, theory creation, deception, mathematics, language, memory, or consciousness—can fundamentally be reduced to the interactions of neurons. If a phenomenon is carried out by the human brain, any valid theory of intelligence must explain how neuronal interactions give rise to such phenomena. This is an absolute requirement that all theories of intelligence must address—no exceptions.

Patternism will solve this problem through patterns. We will meet the challenge halfway by using an intermediary. From the top down, we will show how the products of human intelligence—language, memory, representation, science, and consciousness—can all be reduced to patterns. Once everything is reduced to patterns, the final step is to show how patterns can be reduced to the interactions of neurons—just as, in mathematics, once a theorem is proven, all future proofs need only refer back to the proven theorem.

Patterns serve as our fundamental intermediary, our common ground. They are the universal language through which all complex and intelligent systems can be translated.

We have already solved the first half of the neuronal algorithm problem. Throughout this book, we have reduced all biological and human phenomena to the interactions of patterns. Now, we address the second half. We need only show how neurons can carry out recognition—how they can detect, compare, and store patterns—and we can do so entirely in terms of features.

The Neuronal Algorithm for Universal Recognition:

1. **Detect Patterns** – A pattern consists of features, with consistency being the most crucial aspect. The tools of pattern recognition—*Prediction, Repetition, Natural Selection,* and *Categorization*—are used to detect when this consistency occurs. In neurons, prediction operates through threshold potential, while repetition reinforces frequently activated connections. Natural selection algorithms apply, using prediction accuracy as a selection factor. Categorization can result from pre-labeled datasets or learning from an expert, such as through reading or listening.

2. **Compare Patterns** – Neurons identify similarities and differences between features using the threshold potential mechanism.

   This process begins when the first entity inputs into the neural network, decomposing into a set of features that partially activate specific neurons. A second entity then inputs into the same network, decomposing into another set of features, activating its own neurons. If the neurons partially activated by the first entity match those activated by the second, their activation surpasses the threshold, fully activating them. Fully activated neurons represent shared features between the entities, while partially activated ones represent differing features. A circuit can then silence fully activated neurons representing shared features, allowing the network to focus on partially activated neurons (the differing features) for further analysis.

   This process mirrors how we compare objects in daily life: by placing them side by side and shifting our focus between them to describe their similarities and differences. Similarly, structured data presented in rows and columns facilitates comparison.

V. A Viable Set: Recognition

3. **Store Patterns** – The structure of neural networks inherently forms sets and subsets. With approximately 100 billion neurons in the human brain at birth, no new neurons are necessary—the desired recognition structures technically already exist but are initially hidden. The challenge lies in establishing the correct connections between neurons. Starting with random connections, interaction with the world and the application of pattern recognition mechanisms trains the network, filtering and linking appropriate features to form desired patterns. The separation and groupings of neurons will align with the separation and groupings of entities in reality, enabling the universe to be represented by neurons.

The brain does not operate in isolation. The 100 billion neurons and their connections within a single brain do not limit the number of patterns that can be stored, as human brains are interconnected through communication. Each brain can specialize in specific recognition structures. For example, one brain may specialize in martial arts, holding all relevant patterns, while another focuses on mathematics. Similarly, there are experts in astrophysics, zoology, carpentry, and countless other fields. Rather than relying on the neurons of a single brain, we have access to 8 billion brains, each with its own vast network of neurons. When a specific pattern is needed, an expert can be consulted to retrieve and apply the relevant patterns.

Any medium that embodies the structure of *Universal Recognition* can communicate with another and be translated across different mediums.

Language has this structure. Language can store patterns and their recognition. Neural networks can access the patterns stored in language.

We read—when we need to learn about a subject, we go to a library to find a book on that topic. And we write—by interacting with patterns and documenting the results, we convert the patterns in our neural networks into patterns in language. This allows us to preserve experiences in books, enabling others to learn from them long after we are gone.

We have discussed the neuronal algorithm and the linguistic algorithm, but there is another equally important algorithm: the *Genetic Algorithm.*

The challenge lies in determining the limits of what can and cannot be stored and represented as genes. The human brain, along with sensory organs like the eyes and ears, can be encoded as sequences of (A) adenine, (T) thymine, (G) guanine, and (C) cytosine in DNA. Traits such as camouflage, prey recognition, mate selection, and species interactions are also stored genetically. Additionally, various body morphologies and behaviors—whether flight, swimming, or even consciousness—any traits that manifest without external learning or physical alteration of the organism, are ultimately encoded in sequences of A, T, G, and C.

The task is to find the limits and the connections between an organism's *phenotype* (observable traits) and its genetic representation, the *genotype.*

While codons, the genetic code, RNA, ribosomes, and the processes of transcription and translation have been thoroughly mapped, current evolutionary theory addresses only half of the puzzle: explaining how genes are translated into proteins.

But modern genetics stops there.

What remains unclear is how proteins develop into the phenotype—the human brain, beak length, camouflage coloring, flagella, or sensory organs. Proteins interact with other proteins as well as with the internal and external environments of the organism. This process is far too complex, involving countless interactions, making it extremely difficult to trace every step.

Consequently, biologists often treat these processes as a *"black box,"* focusing solely on the phenotype associated with a given genotype—the inputs and outputs—while ignoring the complex interactions in between.

But why is this allowed? What is it about the universe that lets us safely ignore the middle steps?

V. A Viable Set: Recognition

We can reframe the question and ask: "*When* is this allowed?" When can we treat all the complexity in the middle as a black box?

Again, patterns are the key. Consistency is the answer.

If a given genotype consistently results in a specific phenotype, then all the steps in between must also be consistent. This principle allows the black box approach. It doesn't matter whether there are a dozen or a million steps in between—if the outcome is consistent, the steps must also be consistent. This consistency completes the second half of the *Genetic Algorithm*, tracing the path from proteins to the resulting phenotype. If the steps inside the black box were entirely random, the phenotype would also be random, and the black box method wouldn't work.

The limit is patterns.

What can be stored as genes using the *Genetic Algorithm* are patterns. The *Genetic Algorithm* detects patterns through the process of natural selection. It compares patterns through the competition between living organisms in the struggle to maintain existence. It stores patterns within genes in the structure of DNA, which has its own built-in mechanism for reproduction.

The genetic library uses the medium of genes, the neuronal library uses neurons, and the linguistic library uses written language. All of these libraries exist within the same physical reality, each aiming to describe and store patterns about the same universe. Thus, they can interact and be translated from one form to another.

The neuronal library accesses the linguistic library through reading and writing. This interaction now extends to include access to the genetic library. The genetic library has been translated into language through scientific discoveries and genetic sequencing. Historically, humans—the neuronal library—accessed the genetic library indirectly through selective breeding in animal husbandry and agriculture. In the modern era, direct access is now possible through genetic modification.

All libraries are fundamentally the same.

The genetic library gave rise to the neuronal library. The neuronal library gave rise to the linguistic library. The linguistic library, in turn, led to the most modern form of *Universal Recognition*, manifested through silicon transistors and interconnected computers, creating the cyber library—the internet.

None of these libraries exist in isolation. They interact to form the ***Infinite Library***—a library that encompasses all libraries, past, present, and future, storing every pattern that can possibly exist in this universe.

The *Infinite Library* is fundamentally rooted in biology.

In totality, it is biological organisms that gave birth to the infinite library. Human beings, as biological entities, have harnessed neuronal, linguistic, genetic, and cyber libraries in our relentless pursuit to store and preserve all the patterns we have uncovered. Consequently, a radical redefinition of living organisms is now required. Living organisms must be viewed as collections of recognition agents, localized at a single point. They must be viewed as ***biological libraries***—the sum total of multiple layers of libraries—constantly discovering and storing patterns that help propagate the biological libraries through time and space.

The brain does not exist in a vacuum. Books do not serve their purpose without being read. Genes cannot function without the translation machinery that converts them into proteins. While we have defined the structure for *Universal Recognition*—the framework that stores patterns—this structure must exist within physical reality. It must be connected to an agent capable of performing the act of recognition.

Recognition must occur through an agent.

We define ***an agent as any entity that can act***. It is any entity that can take in features as inputs, process those features, and produce an action as output. A human being is an agent, taking in features through the five senses and producing actions—whether physical, such as operating a tool; verbal, such as naming and identifying; or internal, such as activating specific neural circuits.

A single neuron is an agent, gathering features through its dendrites and producing a signal down its axon in response. An enzyme functions as a reactive agent, detecting features through its molecular structure and facilitating chemical reactions. A virus senses features in its environment via its capsid shell, injecting genetic material into a host cell when a specific set of features is detected. Similarly, a smoke alarm acts as an agent by detecting the features of smoke and triggering a loud ringing sound.

Any recognition agent can be broken down into three essential components: feature inputs, feature processing, and action output.

Components of a Recognizing Agent:

1. **Feature Inputs** – Features are introduced to an agent through sensors. Inputs can range from simple binary signals, such as detecting the presence or absence of light, to more complex variations, like measuring brightness along a spectrum. Inputs may also involve multiple features with varying intensities, as demonstrated by the human eye, which detects color through the relative activation of L, M, and S cone cells. Sensors may also function as an array, where the position of an activated sensor becomes a feature—such as in vision systems, where the grid location of light intensity contributes to image processing. Moreover, features can be internal, with sensors detecting changes within the agent's own body.

Feature inputs do not need to be direct. Through representation, features can be transmitted across time and space, as seen when data is collected and represented in tables, graphs, or as variables in formulas. Features may also pass along chains of representation. For example, signals from activated cone cells in the eye must travel through the optic nerve to reach the brain for processing. As long as each link in this chain preserves consistency and synchronization, the representation remains valid, ensuring accurate feature processing.

However, there are limits to how many features an agent can process at any given time, determined by factors like the number and type of sensors, memory, computational capacity, and the agent's current state. Additionally, agents face bandwidth limitations, as only a fraction of the available environmental features can be detected at a given moment.

Thus, agents must learn to efficiently detect relevant features and filter out irrelevant ones. This involves focusing sensors on specific directions or areas and refining sensor sensitivity to rapidly respond to critical features.

2. **Feature Processing** – Once features are introduced into the system, they must be processed to determine an appropriate action. Feature processing involves various methods, including algorithms, conditional logic, physical mechanisms, or even randomness tempered by evolutionary selection or prediction accuracy. The medium for processing may include silicon transistors, mathematical formulas, biomolecular structures, or networks of neurons.

Feature processing is fundamentally pattern processing. Within the framework of *Universal Recognition*, processing can be viewed as detecting features up to a certain threshold. Once this threshold is reached and enough features are detected, an action is triggered, resulting in recognition.

Threshold systems can vary in complexity. They can be strict, requiring all features of a pattern to be detected before recognition, or fuzzy, where only a subset of features is necessary. Features may have different weights—some are more important and contribute more heavily to recognition, while others may inhibit recognition. Features can interact, form sub-features, or create feedback loops. Temporal factors also play a role, with features decaying over time or new features being introduced.

Advanced processing involves selecting a pattern from a list of learned patterns to match the current situation. This requires the agent to weigh competing patterns, manage interference, and respond to partial or uncertain information. Such complexity can lead to phenomena like illusions or misrecognition.

V. A Viable Set: Recognition

Processing also extends to novel situations. When encountering something entirely new, the agent recognizes the novelty by identifying the absence of a stored pattern. It then initiates learning, seeking defining features through comparisons with previously learned patterns. If no defining feature is found, the agent creates one, staying alert to identify emerging patterns.

Despite its complexity, feature processing can be managed using the tools of Pattern Recognition—*Prediction*, *Repetition*, *Natural Selection*, and *Categorization*—which enable intelligent agents to learn, adapt, and refine their processing systems.

3. **Action Output** – The agent's response triggered after feature processing recognizes a pattern.

The action can take many forms. It might involve injecting genetic material into a host cell, snapping shut the lobes of a Venus flytrap, or activating muscles to flee or attack. The action could be a decision, such as choosing to eat or reject something. It might trigger an alarm, like a security system's alert or the release of bee pheromones to summon a swarm. Alternatively, the action could involve storing features in memory, generating predictions, or creating simulations. In some cases, the action may produce a specific word—an act of identification and categorization through language. The recognition itself—such as the activation of a single neuron—constitutes the action.

Actions may be preprogrammed, unfolding automatically once triggered, or dynamic, adjusting based on the agent's internal state or new input features. Some actions involve passing features to other systems or agents, while others prompt learning responses. In all cases, the action completes the loop of recognition, ensuring the agent can respond effectively to its environment.

There are no inherent limitations to what can qualify as an agent performing the act of recognition. Any entity capable of taking in features through sensors, processing those features consistently, and producing a differentiated action qualifies as an agent engaged in recognition—regardless of the simplicity of its sensors, processing, or actions.

Nothing prevents the combination of distinct recognition agents—each trained to detect specific features, process particular patterns, and produce unique actions—into a single, interacting system of agents that can itself be treated as a single, unified recognizing agent.

For example, we could create a recognition system specialized in external features, comprising distinct agents for audio, visual, and tactile inputs. The outputs of these agents could then be integrated by a higher-order recognizing agent, resulting in a comprehensive and cohesive recognition system.

Similarly, internal features could be managed by specialized agents. One agent might detect damage to other agents, reallocating processing resources to focus on the damage location. Another might monitor the recognition activity of other agents, while yet another detects internal changes within those agents. Some agents could specialize in recognizing novelty, verbalizing clusters of active features, or identifying differences and similarities between two input events. A *"prediction checker"* agent might evaluate the accuracy of stored predictions by comparing them with current inputs, while a goal-recognition agent could match features to a predefined goal, measuring progress by the degree of alignment.

An overarching agent could oversee the entire system, recognizing patterns within it. This supervisory agent would detect when the system is stuck in a loop—repeatedly processing the same features, producing the same actions, and failing to achieve success. Upon identifying such a cycle, the agent would intervene by introducing new features, altering processing methods, or attempting alternative actions to break the loop.

In this way, neurons can be viewed as individual recognizing agents connected to other neurons, forming a network of interacting agents. These neural networks are linked to sensory organs such as the eyes, ears, and skin, and are connected to muscles for movement. Together, they form a single body—a unified recognizing agent. This principle applies not only to human beings but to all multicellular organisms, where systems of interacting recognition agents are built from the fundamental unit of a single agent that detects features, processes them, and produces an action output.

For the overall agent to function effectively, it must remain localized to a single point of coherence. The modulation of the minor recognition agents that collectively form the overall agent is crucial. Excessive interruptions from minor agents can reduce efficiency. For example, interference between sensory agents during feature input might disrupt focus, while interference during feature processing could slow or even paralyze the agent, preventing timely action. Interruptions during action generation could waste energy, especially if an action is halted mid-execution, though such interruptions may be justified in critical scenarios like recognizing pain or detecting damage.

Ultimately, consensus must be reached among the interacting minor agents—either through coordination or inhibition—to ensure that only one coherent set of actions is executed at a time. This creates a singular point of focus, enabling the overall agent to work toward the most essential goal in the given moment. A sense of unity emerges within the overall agent through the seamless modulation of feature input, processing, and action output. This unity is tied to efficiency in time, energy, and effort, making it a trait subject to natural selection and evolutionary refinement.

There is a ladder that must be climbed.

What separates Patternism from all other theories of intelligence is the pursuit of an explanation for all biological phenomena. With recognition clearly defined, we now have the tools to explain the phenomena of sensors, whether biological or mechanical.

A sensor is anything that can detect patterns. Since a pattern is composed of features, a sensor is any physical mechanism that can consistently differentiate between two events.

We begin by asking: "What can be sensed?" What is it about our universe that makes sensing possible?

Sensors exist because of the *Differentiation of Outcomes*. In our universe, consistent interactions between identical entities produce the same outcomes, while interactions between different entities yield consistently different outcomes. In essence, what can be *"sensed"* is defined by whether it follows a regular pattern. If regularity exists, a sensor can be built to detect it. The challenge lies in identifying the right entities and interactions to integrate into a recognizing agent, enabling it to differentiate and reliably detect features.

How good is good enough? A sensor doesn't need to be perfect; it only needs to perform better than random guessing. The key question is whether incorporating a particular interaction benefits the agent. If the benefit outweighs the cost compared to agents without it, evolution will drive the development of more effective and efficient sensors.

The phenomenon of sensors is extensive. All humans have eyes and ears, but what does it mean to see or hear? What about sub-senses, such as detecting light and dark, or distinguishing between sweet, sour, or umami, or even between specific fruits? These nuances extend across all organisms, each possessing various sensors. Some have feelers or antennae, others sense vibrations through silk webs, and still others use chemical sensors to detect pheromones or even magnetic sensors to determine north from south, enabling navigation during migration.

On a microscopic level, protein receptors act as sensors, allowing specific molecules to enter a cell while blocking others. Enzymes sense the correct reactants to facilitate chemical reactions. Beyond biology, humans have developed modern sensors such as thermometers, night vision devices, carbon dioxide detectors, and even locks that detect specific key features.

For a Patternist, **to sense is to recognize**. All sensors function as units of recognition and can be viewed as recognizing agents. All sensors can be broken down into overall agents and sub-agents, as well as into features and sub-features. For example, the eye can be broken down into individual cone cells that detect photons at specific wavelengths, just as the ear can be broken down into individual hair cells of varying lengths that respond to specific vibration frequencies. The result is a spectrum created from numerous individual units, each consistently distinguishing between two events.

V. A Viable Set: Recognition

For a Patternist, *to sense is to represent*. Night vision optics translate darkness into varying shades of green, while thermal vision converts heat into gradients of white. When we look at a vehicle dashboard, we encounter various representations of sensors that inform us about the vehicle's status and any changes that occur. We have a fuel level sensor, depicted as a gas pump icon with a dial from empty to full. A deflation sensor lights up for flat tires, and an engine temperature sensor displays a dial from hot to cold, alongside sensors tracking RPM and speed.

As long as sensors remain consistent and maintain synchrony as a representation, an agent can detect new features, differentiate them, and incorporate them into its processing to determine the appropriate action. This principle applies universally—from operating vehicles to interpreting readings from instruments, tools, or experimental data across any field or context.

Fundamentally, sensing is the act of categorizing, labeling, naming, and identifying, as they all share the same underlying root. To sense is to recognize. To sense is to represent. To sense is to detect patterns.

There is no reason to waste time on abstract thought experiments when there are billions of real-life examples to observe and learn from. Why debate the nature of color-blind Mary and the color *"blue"* or argue about the nature of *qualia*, when we cannot yet explain how a virus identifies a host cell, how a bat uses echolocation, or how we interpret readings from instruments?

A Patternist prioritizes practicality. When it comes to sensors, the key questions are: "Can it differentiate entities?" Can it separate or group them? Can it categorize reliably and make accurate predictions?

Is there a pattern?

There is no isolation.

Agents interact and compete with other agents. Agents deceive other agents.

Deception occurs when one agent manipulates the recognition system of another. Since recognition relies on detecting features, once we understand the features a target agent's recognition system act on, we can exploit that system by either artificially triggering those features beyond the detection threshold or inhibiting them to blind the target agent, effectively controlling its response.

For example, a lure that mimics a minnow or fly can trick a fish into biting a hook. Migrating birds can be deceived by decoys resembling other birds. A magician manipulates the audience's sense of object permanence, making coins and cards seem to vanish and reappear, creating surprise and delight. Optical illusions exploit how the human eye processes visual information. On a microscopic level, pharmacology—whether through drugs, antidotes, poisons, or artificial sweeteners—operates by tricking cell receptors, either forcing them to open or blocking activation.

Deception can also be passive. We can fool a target agent by preventing its recognition system from triggering a response. If a target agent detects motion, we can remain still and only move when it's not looking. If it relies on scent, we can mask our odors or stay downwind.

Agents cannot afford to overreact. Constant vigilance wastes energy and increases stress. Thus, agents learn to ignore unimportant stimuli, such as background noise or visual clutter. By sharing features with the environment, intelligent agents can avoid detection, giving rise to camouflage. From octopuses blending with coral by changing their color and texture, to orchid mantises mimicking bright flowers, camouflage is widespread in nature. Modern military uniforms adopt this principle with patterns that adapt to desert, forest, or snow environments.

Intelligence is the ability to find and utilize patterns. A truly intelligent agent will seek to exploit the patterns found in other agents. It will learn to deceive other agents. Human beings are highly intelligent. Through interactions and competition with equally intelligent agents, we have become expert deceivers.

We create propaganda. We control media narratives, edit photos and videos, and manipulate language to shape perceptions, influence behavior, and steer public opinion. If the target holds preexisting beliefs, we reinforce them. We repeat lies until they unconsciously become accepted without question. We use word associations, speak

V. A Viable Set: Recognition

with confidence, wear symbols of authority like lab coats, display diplomas, and attach the word *"science"* to dubious fields to lend credibility. We manipulate statistics, confuse with numbers, and embed half-truths within truths. We predict outcomes we already know to establish trust when they come to pass.

All for one purpose: to control what the target agent recognizes as *true* or *false*, shaping their beliefs and actions.

We craft narratives. We learn the grand narratives—the myths a civilization follows—and adjust our deceptions to fit them. We exploit archetypes, casting rival leaders as villains and creating stories to reinforce that label. We call them dictators, selectively quoting and editing their words out of context.

We find the opposition and cast them as heroes and freedom fighters. We show images of women and children, invoke fear, and feed the narrative with false stories and modified history. In doing so, we craft grand narratives of good versus evil, deceiving entire populations.

And when all else fails—we ignore, censor, and erase. We control the channels that provide information to the target agent. We control what is fed into the target. We control the sources: television, magazines, *"trusted"* media. We dominate the mainstream and centralize it, remaining ever-vigilant for emerging sources that challenge the prevailing narrative or present alternate views of reality—and we cut them down.

And we forget. The injustices carried out by our own hands, our own nation, our own lies are washed away with the passage of time and the deliberate avoidance of our own eyes in a final act of self-deception.

There are no limitations. There is no isolation. There is only Patternism.

Patternism reduces in order to build and create.

By breaking down recognition into features and patterns, new forms of recognition systems can be constructed. Ask: "What is the fundamental unit of recognition that defines an agent?" What features does the agent receive, how does it process them, and what actions result? What constitutes the overall unified recognition system? What is the network structure, and how do agents interact within the system? Is there modulation, and how does the overall recognition system decide on a final course of action?

New types of sensors can be developed, enabling new ways to detect, categorize, and identify phenomena. Ask: "Is there a separation between entities and events?" Is this separation regular, and can it be integrated into a sensor? Can representation be synchronized to where the agent can effectively use the sensor?

The limits of recognition can be explored. Ask: "Is the recognition system too sensitive or too strict in feature matching?" When is it better to risk a false positive, and when is it better to be overly strict, letting opportunities pass?

Under what conditions is recognition possible? If entities have been described, labeled, and categorized by one intelligent agent with consistency, shouldn't it be possible to replicate this recognition in another system? Can the same recognition processes be copied and implemented in machines?

New methods of deception will emerge. Ask: "What patterns does the target agent rely on?" What sensors are available to it? By analyzing the specific features the target agent's recognition system operates on and understanding how it processes them, we can exploit and control the agent.

To create intelligent agents, we ask: "How does an agent differentiate and categorize?" How does it identify similarities and differences between entities? How can it recognize new patterns and utilize them?

And we can create conscious agents by asking: "Can an agent recognize itself?" What specific features separate an agent from other entities, and how can the agent detect and process those features? Can the agent recognize external events caused by its own actions and identify what it can control? Can it trace an event back to itself? What features reveal to the agent that it caused the event?

Can the agent recognize the patterns and features its own recognition system operates on, break those patterns, and thus change

V. A Viable Set: Recognition

the way its recognition system inputs features, processes features, and carries out actions? Can the agent recognize patterns in its own behavior and break away from them?

Can an agent create an ideal pattern of itself and work toward becoming that ideal?

Can it recognize whether it has succeeded in becoming its ideal self or if it has failed—and that it is out of time?

Can an agent be made to suffer?

Principle 3. Reproduction

Reproduction is the propagation of patterns through time and space.

The phenomenon of reproduction is extensive, encompassing processes at every level of biological complexity.

At the molecular level, mitosis enables a strand of DNA to split and reform into two complete daughter strands. Under the right conditions, individual protein structures can self-replicate, as seen in prions. Viruses reproduce by hijacking other cells translation machinery. Symbiotic organelles like mitochondria replicate independently within host cells, forming populations that are distributed between two daughter cells during cell division.

Reproduction can be deceptive and complex, often making it difficult to pinpoint where and in what form it occurs. For instance, in meiosis, DNA is reproduced, but the resulting daughter cells are intentionally made genetically distinct, each containing only one set of chromosomes, unlike the two sets produced in mitosis.

In multicellular organisms, every cell reproduces by dividing into two daughter cells. However, only germ line cells—the single fertilized egg and sperm—carry life forward, while somatic cells perish along with the organism. Some cells, like cancer cells, deviate from this pattern, dividing continuously in a way that ultimately harms the organism.

In sexual reproduction, two concurrent forms of an organism—male and female—must interact to produce offspring. In some species, sexual dimorphism is extreme. For example, the male of certain anglerfish species exists primarily as genetic messengers, their sole purpose being to deliver genetic material.

Some organisms bypass sexual reproduction entirely, cloning themselves by breaking off a section and regenerating, as starfish and many plants do. Others adapt their reproductive strategies to environmental conditions, reproducing asexually in stable environments but switching to sexual reproduction when variation is needed to adapt to a changing environment.

Reproduction also occurs at the level of biological colonies. For example, an ant colony reproduces even though only the queen lays eggs, while all the individual workers are sterile. This principle applies to other eusocial organisms like bees and naked mole rats. Similarly, multicellular organisms can be seen as colonies of individual cells working together.

Individuals die, but the colony remains. Penguin colonies, meerkat clans, and chimpanzee communities. In human societies—family, tribes, and nations—each persist through time, despite the continual replacement of members, leaders, and institutions. Similarly, sports teams, universities, and armies maintain their identity and function, even as their constituents come and go.

What, then, truly reproduces? Is it solely the genes? The machinery that translates genes into proteins? The proteins encoded by those genes? Perhaps it is the physical traits expressed in the complete organism? Or is it the entire colony—the species—that reproduces?

For a Patternist, the totality of all levels of reproduction and their outcomes must be considered. Every level—genes, proteins, physical traits, organisms, and colonies—participates in the process of reproduction. These levels are intimately connected, each supporting and reinforcing the others, creating an interdependent system of continuity.

There are no limitations. The continuity extends beyond the biological domain. It extends beyond the colonial level and covers what the organisms themselves can craft and learn. The principles underlying reproduction are not confined to living organisms but are evident in other systems that rely on replication and continuity.

Manufacturing and industrial production are included: the reproduction of arrowheads, cans of cola, cars, smartphones, screws, and hammers. When we account for the machines, factories, blueprints, and assembly lines involved, we find that many processes used in the reproduction of these items are comparable to those found in the biological domain.

Reproduction is not limited to physical objects; it also includes the reproduction of techniques. This covers the preparation of a specific dish using a family recipe, the passing down of effective moves in martial arts from master to student, and the practice of variolation—a form of smallpox inoculation introduced to Europe after being observed in the Ottoman Empire. All schools can be viewed as systems for reproducing techniques, with teachers, coaches, textbooks, instruction manuals, and practice all contributing to this process.

Reproduction includes intangibles. The reproduction of the *Mona Lisa* on a poster or t-shirt. The reproduction of a novel through hand copying, translation, or digital duplication. The adaptation of a story across different media, including remakes with new actors playing the same characters. The copying of essential documents, such as birth certificates or driver's licenses, and the replication of a computer virus.

Ideas reproduce. Theories reproduce. Beliefs reproduce. They replicate within human minds, often with significant consequences.

It begins with a single idea, originating in the mind of an individual. The idea finds utility, holds some reflection of reality and predictive power, or serves as a tool to gain influence by appealing to emotions, human nature, and the spirit of the times. It becomes viral. The idea spreads through communication—lectures, books, and media. It is given a name, identified as an "-*ism*." The idea spreads and reproduces in other minds. Over time, it is refined and becomes doctrine. As doctrine, it seeks to dominate and eliminate all other ideas. There can be no competition, no rivals. Doctrine becomes ideology.

The medium does not matter.

With patterns at the core, reproduction can occur in any medium, adapting to the systems it inhabits. Anything capable of encoding and transmitting a pattern is participating in reproduction. Biological cells replicate through DNA, factories reproduce products through assembly lines, stories are retold across generations, and ideas spread from mind to mind. Regardless of the medium—organic, mechanical, digital, or conceptual—the essence of reproduction remains the same: to preserve, transmit, and propagate a pattern.

V. A Viable Set: Reproduction

In this way, reproduction transcends physical boundaries. Though the medium may change, the drive to reproduce patterns is a universal constant. Patterns connect the non-living to the living, the physical to the abstract, allowing each to carry forward the structures that define their existence.

Reproduction, therefore, allows entities to transition between different mediums, moving fluidly from one form to another.

A machine can become a man. The pattern that constitutes intelligence can transition from the biological medium of neural networks to the digital medium of artificial networks. When the principles of intelligence are reproduced within a machine, the machine itself is transformed, becoming capable of human-like recognition, understanding, and adaptation.

In the same manner, a man can give birth to an idea, transcending his physical form to become the idea itself. The idea becomes recognized by the same name as the man who conceived it. Thus, a man can become an idea.

An idea serves as the foundation of an internal universe, a construct that takes root in the minds of others, shaping their thoughts and guiding their actions. It can spread, multiply, and influence countless lives, propagating itself as a distinct form of existence. Thus, an idea can become a god. As long as there are adherents, an idea can endure for millennia, transcending its origin.

The connection is clear—from man to idea, to a sacred belief.

And thus, a man can become a god.

Have men not become gods throughout history? Are there not countless instances where a man is transformed into something revered, immortalized, and worshiped?

Did Augustus not become the deified Emperor Augustus, worshiped in temples long after his death? Did Siddhartha Gautama not become the Buddha, with his teachings on transcending suffering still followed by millions today? Does the Capitol Building in Washington, D.C., not bear the fresco *The Apotheosis of Washington*, depicting George Washington ascending to the heavens as a divine figure?

Patternism

Great men, founders of nations and architects of civilizations—they become more than human. Their statues stand tall in public squares, their portraits hang in homes and institutions, their words and deeds echo through generations.

Their patterns—their ideas, values, and actions—are preserved, reproduced, and propagated across time, transforming them from mere mortals into enduring symbols of reverence.

Is apotheosis not, at its core, a biological phenomenon?

Just as genes replicate through generations, so too do the patterns of great individuals—through stories, symbols, and collective memory. The drive to immortalize and replicate transcendent figures mirrors the biological imperative to reproduce and endure.

In this sense, apotheosis is not only a cultural or spiritual event but a biological one—a testament to the power of patterns to transcend mediums, from flesh to idea, from life to legend.

If it can be reproduced, it is a pattern. If it is a pattern, it will transcend medium.

Intelligence is a reproducible pattern.

Through Patternism, intelligence will be reproduced not only in machines but in any medium capable of encoding and transmitting patterns. Patternism is the theory—the blueprint—that will give birth to forms of intelligence that transcend biology. Intelligence will be elevated beyond the human medium, beyond the biological, enabling it to inhabit synthetic systems, digital frameworks, and even domains yet to be imagined.

Consciousness itself is a reproducible pattern.

Through Patternism, consciousness will transcend humanity, inhabiting any medium capable of supporting the structure of consciousness.

To think, to be, and to know. Through Patternism, intelligence and consciousness will no longer be bound to flesh and neurons but to the patterns that define them.

Reproduction is the propagation of patterns through time and space.

V. A Viable Set: Reproduction

But how does a pattern begin? What is the origin that initiates a pattern? If we are to observe reproduction in absolute terms—adhering strictly to the principle of no isolation, where all reproductive phenomena exist within the same physical universe with no separation between reproductive acts—then, in theory, we can trace all acts of reproduction to a single origin.

All books, technologies, and manufactured goods can be traced back to human beings, who are biological organisms, subject to reproduction. Through evolution, the lineage of human beings can be traced back to a single-celled organism that existed billions of years ago.

But how did this single-celled organism arise? At this point, the lineage is unclear, and the best we can do is hypothesize that, in the very beginning, there existed a self-propagating protein structure that eventually gave rise to the first single-celled organisms. But how did this self-propagating protein structure originate? What is needed is a ***Prime Origin*** that describes, in completely probabilistic terms, how reproduction—the act of self-propagation—begins from scratch.

For a Patternist, the phenomenon of reproduction is guaranteed by the inherent properties of the universe. If the possibility of reproduction exists, no matter how small the probability, given enough time and randomness, the state in which reproduction occurs will eventually be achieved.

Patternism

Prime Origin: Conditions Necessary for Spontaneous Reproduction:

1. **Reproductive Possibility** – The existence of a state within the system that includes entities capable of reproduction. The probability of reaching that state may be minimal, but as long as it exists, it remains achievable over time.

2. **Inherent Randomness** – The system can transition through various states and events; it is dynamic rather than static.

3. **Passage of Time** – Time allows a system to change states, essentially *"rolling the dice."* With the passage of time, the system will cycle through increasing numbers of potential states.

4. **Resource Limits** – There must be sufficient resources for the reproducing entity to recreate and propagate itself. Resources do not need to be infinite; they need only suffice to maintain the population and allow evolutionary processes. Resources may be recycled once the population reaches maximum capacity, as reproducing entities will have access to fellow members, which they can consume and break down as resources through phenomena such as predation and cannibalism, limited only by energy constraints.

With enough time, the state containing self-propagating entities will be achieved. The probability of reaching spontaneous reproduction approaches certainty as more *"rolls of the dice"* occur. All that is required is the emergence of the one entity capable of reproduction—the *Prime Origin*. Once reproduction begins, a population will form, allowing evolution to take over, with self-propagation increasing in complexity and efficiency through evolutionary processes.

The only significant limitation is the occurrence of catastrophic events that might reset the entire process. Even then, for the process to be halted entirely, such an event would need to eliminate all reproductive entities. If even one self-propagating entity survives, a population will eventually reemerge.

For a Patternist, among all possible states in this universe, only the state that contains self-propagation is distinct. The only phenomenon that truly matters is reproduction. All else will be eliminated by the passage of time. This applies universally to all that exists. No exceptions.

Reproduction cannot occur without patterns and the *Paradox of the Pattern*.

All cans of cola are the same, and all cans of cola are different. All smartphones are the same, and all smartphones are different. All cats are the same and all cats are different.

The phenomenon of reproduction utilizes the *Paradox of the Pattern* and makes full use of the *Ship of Theseus* problem.

For a Patternist, the solution to the *Ship of Theseus* problem does not lie in deciding whether the current ship, with all its planks replaced, or the reassembled ship, made from the original planks, is the *"true" Ship of Theseus*. Instead, the core aspect of the *Ship of Theseus* problem is its demonstration of the phenomena of reproduction, along with the processes of consumption and repair, which contribute to reproduction.

Through the *Paradox of the Pattern*, the *Ship of Theseus* problem demonstrates how two ships can emerge from the original. From those two ships, four can be made, then eight, and so forth—just as a single strand of DNA is replicated into two daughter strands, then four, and so on.

This principle applies universally to all reproductive phenomena. The only requirement is that the components a reproducing entity uses to recreate itself must be interchangeable with components in the environment—that is, the components must be a pattern within the environment. In the case of ships, planks of wood are a regular feature in the environment, interchangeable with other planks. For DNA, the nucleobases—guanine, cytosine, adenine, and thymine—are abundant. For living organisms, the component nutrients—proteins, fats, and minerals—are acquired by eating and breaking down other organisms.

For viruses, there are translation machinery in the environment available to be hacked and reprogrammed. For manufactured goods, bolts, screws, and electronic components are similarly interchangeable.

But what determines if two component entities are the same?

What makes one plank of wood the same as another? What makes one adenine molecule the same as another adenine molecule? What makes one capacitor the same as another capacitor, or one bolt the same as another bolt?

Two entities are considered the same if they share the same defining features.

Using the principle of the *Differentiation of Outcomes* we can perform physical tests to determine whether two entities are the same.

In terms of component parts, if an overall entity can replace a component part and still function in the same way as before—determined through testing—then the two components can be said to share the same defining features and can thus be categorized the same. In this case, it is the overall entity that determines which features are important.

For example, for bolts, the defining features may be a specific length, diameter, and thread count. Once a bolt is replaced within an overall entity, such as a door hinge, we can test whether the door still operates as it did before: does it open and close as before, or does it fall from the frame?

The same principle applies to blood transfusion. After blood from a donor is transfused into a patient, does the patient function as before, or do health complications arise due to rejection? Was the donor blood compatible? Were the key features of the donor's blood and the patient's blood the same? What features of blood enable a successful transfusion?

This concept also applies to living organisms. Can two organisms be categorized as the same species? If the genetic material of one organism replaces that of another, will the organism still function? If one organism replaces another in a population, will the population continue to propagate? What, then, are the defining features of a

V. A Viable Set: Reproduction

species? Is it genetic or morphological similarity, or is it simply the ability to reproduce viable offspring?

Just as features can have sub-features, components can have sub-components, and sub-components can have sub-sub-components. The questions of "What are the defining features?" and "Are two entities the same?" apply equally across these levels and can be solved using the function test at various scales. At each level, an overarching entity at the scale above determines the defining features of the current components.

This **Component Part Test**, used to determine if two entities are the same, is a direct application of the *Ship of Theseus* paradox. Like the *Ship of Theseus*, the solution to component part testing is relative to what is taken as the overall entity. In the *Ship of Theseus*, is the overall entity the museum, making the ship merely a component part of the museum? Is the overall entity Theseus and his crew? Is the overall entity the ship itself, if the ship is conscious and has a representation of itself to compare to? Or is the overall entity a philosopher, with the *Ship of Theseus* as a component of a thought experiment? What, exactly, is the overall entity being referenced?

Multiple overall entities give rise to multiple sets of defining features. How to group and define entities becomes difficult because there is not one single valid means, but multiple valid means to define features depending on the overall entity in question. This problem of relativity in defining the overall entity is universal and lies at the heart of the *Species Problem*: how and when to define species.

The Species Problem is a reframing of The *Ship of Theseus* problem. Patternism recognizes that the difficulty in defining species is the same as the difficulty in identifying the true *Ship of Theseus*. In defining species, what is the overall entity to which classifications are made? Is it the director of a museum, who must organize preserved specimens in the museum's collection? Is it the geneticist, who sequences genetic material and categorizes species based on genetic patterns? Or is it a medical doctor, who must classify infections—both viral and bacterial—based on symptoms and treatments? What is the overall entity being referenced?

While defining features may vary depending on the overall entity considered, this does not mean feature determination is entirely relative. There is a final, objective feature determiner. When taken to the extreme, the ultimate determiner is the universe itself, as all that exists are component parts of the universe. No exceptions.

The defining features the universe selects for is what is not eliminated. What is not eliminated by predation or rival agents. What is not eliminated by floods, fire, or asteroid impacts. And what is not eliminated by time. All that remains are entities capable of continuous reproduction. Reproduction is the ultimate defining feature. The universe will eliminate all else.

Reproduction gives rise to entities that are the same. Reproduction gives rise to patterns. But there is a paradox to patterns—the same entities can be different. Thus, reproduction gives rise to entities that are both similar and different. In practice, there will always be variation among entities that are categorized as the same.

This variation arises naturally. The key question is whether the variation impacts the defining features, and if so, to what extent. What are the allowable tolerances in defining features that still enable the overall entity to function, and what are the cutoff limits beyond which functionality is lost?

The *Component Part Test* can be used to determine the allowable tolerance and cutoff limits by varying the component part and then subjecting the overall entity to a function test. In manufacturing, engineering blueprints specify the acceptable limits of critical features of a part. After production, inspectors measure the part to ensure it falls within these tolerances and may also test it for fit and functionality by assembling it into the larger system. In living organisms, genetic variation occurs naturally, and the tolerance and limits are ultimately determined by the organism's ability to survive and reproduce.

The emergence of variation among entities categorized as the same is not inherently negative. In the negative case, extreme variation can lead to damage or a loss of function for the overall entity, necessitating rejection of the component parts. But in the positive case, variation can enhance the overall entity, making it more efficient or even enabling new functions. This can be seen in the upgrading of a vehicle or a personal computer. In biology, if we view an organism's components as protein structures, variation in these proteins may lead to improved functionality for the organism. Going further, at the level of protein sub-components—genetic sequences—variations may produce more effective proteins, enabling evolution to progress through these microscopic changes.

We now return to the *Ship of Theseus* paradox and apply variation. We introduce variation to the component parts—the planks of wood that form each ship.

We start with a population of *Ships of Theseus*. Through the replication process, from one ship we build two. From two, we build four. From four, we build eight, and so on, with each new ship being assembled on the deck of the old ship by a robot shipbuilder.

Now we ask, "What if, in the process of building new ships, the shipbuilder makes a rare mistake?" For example, instead of using a brown plank in a particular position, the shipbuilder uses a white plank. Additionally, each shipbuilder has limited access to information: they are assigned to a single ship and can only observe their own ship as a guide for plank placement on the next ship to be built.

The result of this process is a lineage of *Ships of Theseus* inheriting unique variations that differentiate them from other lineages—variations that accumulate over time as more generations of ships are built. After a hundred generations, a thousand, or even a million generations, we ask, "Is this the same *Ship of Theseus* as it was in the beginning?"

All *Ships of Theseus* are the same, and all *Ships of Theseus* are different.

From the population of *Ships of Theseus*, a group sails off to North America, another sails to Australia, and a third remains in Greece.

Each ship is assigned a single robot shipbuilder. There is no land, so all new ships must be built on the deck of an old ship by a shipbuilder. Now, however, the shipbuilders can communicate and observe nearby ships as they build new ones. And, as before, shipbuilders can make mistakes.

The result will be inherited variations, but now these variations can transfer within the local population, with isolated populations developing their own unique sets of variations.

The variations must not harm the ships. Sunken ships do not reproduce. Sunken ships do not have decks on which a shipbuilder can build new ships. Sunken ships cannot be observed by other shipbuilders when constructing new ships. The *Ships of Theseus* must at least remain functional.

Add in different local conditions: North America may be colder, requiring a thicker hull to help break through ice, while Australia, with its shallow and tidal waters, may need a shallow-draft keel to navigate through coral reefs and low waters. Add that each shipbuilder can build more than one ship depending on the quantity of components they can gather aboard their assigned ship. Add in that some variations can be beneficial in helping the shipbuilder gather more components depending on the local environment.

What we have, as a result, is the process of evolution applied to the *Ships of Theseus*. But more importantly, what we have truly just derived is the process of *evolution* from the *Ship of Theseus* paradox. If philosophers had closely examined the reproductive aspect of the *Ship of Theseus* paradox, a theory capable of explaining the process of evolution would have been found and described a thousand years earlier.

Again, we ask, "What is a *Ship of Theseus*?" Can a ship whose ancestors sailed to North America and adapted over generations to the environment of North America still be called a *Ship of Theseus*? What about a ship whose ancestors sailed to Australia? Can a ship from North America and a ship from Australia even be called the same?

Can a shipbuilder onboard a ship from North America observe a ship from Australia and use it to build a hybrid ship? Will the resulting hybrid ship be functional? What about after a thousand generations, when the ships have evolved to become highly specialized for their specific environmental conditions?

V. A Viable Set: Reproduction

What is a *Ship of Theseus*, and how is it defined?

What is a species, and how is it defined?

What determines if two entities are linked by reproduction? How do we determine if two entities that share similar features are actually related through reproduction and not merely the result of independent development?

What causes independent development in the first place?

If two environments are governed by the same set of rules, face the same set of problems, there will be a ***Convergence of Solutions***—that is, agents in similar environments will independently arrive at similar solutions through random trial and error, without any need for interaction. These solutions can take various forms: a technique, a set of behaviors, a body morphology, a physical trait, or even protein structures. The only requirement is that the solution be consistent, forming a pattern and thus usable as a defining feature.

Convergence of Solutions can be observed in biology through ***Convergent Evolution***, where distinct populations of organisms independently evolve similar features in response to comparable environmental challenges.

For example, eyes, flight, and sexual reproduction have each evolved multiple times independently. Similar body forms are observed in both modern-day horseshoe crabs and extinct trilobites, as well as in extinct ichthyosaurs and modern dolphins. Compare the North American wolf to the Tasmanian wolf. Organisms in cave environments tend to share features like long limbs, blindness, and translucency, yet are more genetically similar to local species outside their caves than to other cave-dwelling organisms in distant locations. Various forms of eusocial behavior have evolved independently in ants, bees, shrimps, and naked mole rats.

In *Convergent Evolution*, interaction between the agents is unnecessary, and their physical environments are usually separated by significant distances in both time and space.

However, *Convergent Evolution* comes with a qualification. Technically, three agents are involved: the two agents sharing similar features and a third agent that defines and observes these shared features. This third observing agent is often a scientist who detects the repetition and patterns through careful observation.

Convergent Evolution is not useful to the two agents sharing similar features; it is only useful to the third observer. But to be truly useful, *Convergent Evolution* must be viewed as a *Convergence of Solutions*. The observer must recognize that the independent evolution of similar features suggests a fundamental set of rules governing both populations. These shared features provide clues—if not the exact solutions—to a common problem.

If a solution works once, it can work again. If it works twice, it can work many times over. If it works across two different mediums, it can work across many different mediums. When the observer can fully describe the fundamental rules defining the problem, they have found a universal solution—a universal pattern.

Ultimately, all agents exist in the same physical universe governed by the same set of physical laws. The final *Convergence of Solutions* is the recognition that all populations of agents, both current and extinct, share a common set of features—a common set of solutions—that, if not inherited, have been independently developed again and again. For a Patternist, this set of solutions exists and is described as the *viable set* of principles: *Representation, Recognition, Reproduction,* and *Randomization.*

Every entity is technically unique. Starting from scratch, every virion is unique. Every cat is unique. Every human being is unique. Individuals must be treated as individuals. A single member of a species is distinct from the species itself. How, then, can any individual entity be linked to another? How can two individual entities come to share features? How do we determine if two entities that share features are linked by reproduction, starting from a place of no prior knowledge?

In absolute terms, it is possible for the universe to produce two similar entities purely by random chance, but this probability is extremely small. For the universe to produce three such similar entities by random chance becomes exponentially smaller. For the universe to produce four becomes exceedingly smaller still.

V. A Viable Set: Reproduction

In purely probabilistic terms, reproduction greatly increases the likelihood that the universe can produce two or more similar entities. Unlike *Convergence of Solutions*, an entity is said to be linked by reproduction to another if the time it takes to develop common features is less than the time required to develop those features independently through trial and error.

What is the timeframe for a brain to learn a solution solely through trial and error? Days, months, or even years, depending on the difficulty of the task and the number of experiments. Now, what is the timeframe for a brain to learn the same solution through a school, a teacher, or a book?

What is the timeframe for an entire species to evolve a solution through trial and error? Tens of thousands of years, if not longer, depending on factors such as mutation rates, population turnover, and population size. Now, what is the timeframe for an entirely different species to acquire the same solution through modern genetic engineering?

What is the timeframe for a company to develop an entirely new product from scratch without observing the market or drawing on previous designs? An automobile from scratch? A smartphone? A CNC lathe? All without access to existing mechanical or electrical components like screws, capacitors, transistors, and so on?

If we encounter two entities with shared features and solutions, we first ask, "What is the probability that each entity developed these features and solutions independently?" What timeframe is involved, and is there a history of experimentation or trial and error for each? If that probability seems unlikely, we can suspect that the two entities are linked through reproduction—that a moment of interaction, past or present, has resulted in both entities sharing features and solutions.

If we find three such entities sharing the same features, our confidence in a reproductive interaction increases. With four, it grows further. The more entities we find with shared features and solutions, the more confident we can be that a reproductive interaction has occurred, as the probability of random convergence diminishes.

Reproduction is the mechanism that allows solutions to pass from one entity to another. However, all solutions are initially discovered through trial and error—the cost of randomization must be paid. The question then becomes: "After investing time and energy to find a solution, is there a way for other entities to acquire that solution without paying the initial cost?" If the answer is *"yes"*, then reproduction has occurred, and the question shifts to how and by what means this reproductive interaction enabled the sharing of solutions.

There is an original, and then there is a reproduction of the original. The reproductive interaction is what links the two. All of the reproductive phenomena discussed earlier are examples of interactions that result in the sharing of features between entities.

In general terms, there is direct reproduction, such as cloning and replication, where one entity directly creates another by gathering and assembling the necessary components. In complex entities capable of independent resource gathering, the reproductive process may not produce the complete entity but only the essential mechanisms for further assembly. The new entity must then gather additional resources to complete its body and eventually gain the ability to reproduce. This is common in most living organisms, where the parent technically only produces a fertilized egg, which matures into an adult organism independently.

There is the reproduction of entire populations. In phenomena like colony formation, a population of reproducing entities that share features is divided, moved to new locations, and allowed to grow. Over time, this process repeats, much like how towns and cities form, sharing the same language and culture, or how a population of mitochondria divides between daughter cells.

Chimeric reproduction occurs when components from different entities are combined to create a single, new entity. Many become one. This new entity shares features with multiple parent entities but is not identical to any of them. It may even acquire new abilities or develop novel features due to the combination of its components.

For example, smartphones are created by combining a phone, touchscreen, and radio into a unified device. Similarly, if a gene is viewed as a component, sexual reproduction can be considered a form of chimeric reproduction. In the case of sickle cell disease, the condition arises when an individual inherits two recessive genes. However, if an individual inherits only one recessive gene, they become resistant to malaria. In sexual reproduction, genes are reshuffled, and male sperm combines with a female egg to produce offspring that share features with both parents while remaining distinct from either.

Similarly, if words are treated as component features learned by a neural network AI, the combination of input words to produce new outputs is analogous to reproduction, even if the output is entirely novel. Whether the neural network is trained on next-word prediction or images, and whether the output is a sentence, artwork, or music, the result is a reproduction that combines learned features in new ways—recombining known patterns to generate something new.

There is a process of reducing an entity into a set of representations, capturing only its essential features, and then regenerating the entity through translation—usually via a manufacturing process where entities are created based on the features detailed in the representation. For example, a fighter jet can be reduced to a collection of blueprints. These blueprints can be contracted out to various external companies, which manufacture specific components according to the designs and ship the completed parts back to the main company for assembly.

In corporate espionage, blueprints can be stolen and used by a rival nation to recreate the original product. In biology, a desired feature of an organism can be reduced to its genes, which are then transferred to another organism through genetic modification, allowing the modified organism to express the desired feature.

In manufacturing, a production facility directly produces entities, ensuring consistency of defining features through quality control and testing. The reproduction of the facility itself must also be considered. When viewed as a whole, the facility must survive and gather its own resource components. The materials needed to make the products have to be physically mined and transported to the facility. Workers must be employed and paid, and as workers quit or retire, they must be replaced. Additionally, the entire facility must be funded through the sale of its products.

There is observing and mimicking, where one student entity observes a teaching entity and mimics it to learn the solution the teaching entity has found. The student will try to replicate any feature displayed by the teaching entity that the student can detect until the same solution is achieved. This applies whether the teaching entity is aware of being copied or not, such as in the development of Velcro, inspired by observing the hooked attachment structures of seeds that allow them to be carried off and dispersed. This also applies to stolen technology, where a captured weapons system, such as a downed drone, fighter jet, or intact missile, is taken apart, observed, and reverse-engineered in a lab.

In the same manner, teaching can be deliberate, as between a teacher and student. The teacher could be a textbook, a video, or direct hands-on experience, with a teacher nearby pointing out which features to focus on and correcting the student if mistakes are made. The student can then be tested for competency by replicating the solution in a lab, during practice, or even in a live setting.

There is counterfeiting, where a reproduction is made with enough features to fool a target agent into triggering a false response. Examples include replicating the features of baitfish in plastic lures to trick fish into striking, crafting decoy ducks and using duck calls for hunting, or producing counterfeit luxury goods with inferior materials but selling them at premium prices. Sometimes, the reproduced features are sufficient to trigger the desired response from a distance, but closer inspection may reveal distinctions between the counterfeit and the original. For instance, a security guard's uniform may resemble a police officer's from afar, and works of art, musical covers, or parodies may mimic well-known characters, melodies, or settings while remaining different enough to avoid copyright infringement.

Conversely, there are reproductive interactions designed to limit or prevent reproduction, such as patent laws. By filing a patent, an inventor safeguards the critical features of a new invention from unauthorized replication. Similarly, trade secrets and hidden features act as barriers to the unauthorized reproduction of component parts.

The interplay between counterfeiting and reproduction-prevention strategies is particularly evident in cases like counterfeit currency. Counterfeiters attempt to replicate currency features with high accuracy to deceive the marketplace. In response, governments incorporate complex features into their currency, often hidden or detectable only with specialized equipment, to make replication difficult. As technology advances, both governments and counterfeiters must continually adapt to each other and to the evolving technological landscape.

All reproductive interactions aim to reproduce entities that share the same features. But reproduction is not solely about quantity. There is a time dimension to account for.

Reproduction is the propagation of patterns through space and time. Entities can be separated by space and compared for similarities, prompting the question, "Do these two entities—existing in different locations—fall under the same category and, thus, can be given the same name?"

Similarly, a single entity can be separated by time and compared to itself, asking, "Does this entity, treated as two different entities existing at two different moments in time, belong to the same category and, therefore, can be given the same name?"

Quantity does not matter. A pattern can propagate with only one lineage.

The *Ship of Theseus* paradox is about the reproductive interaction that is the propagation of a single entity—a single pattern—through space and time. It is the reproductive interaction of replacing all the component parts that make up an entity while still categorizing that entity as the same as it was before the parts were replaced—just as a multicellular organism continues to exist as a single entity, even though each individual cell eventually dies and is replaced.

The propagation of one. A body of one. The act of consuming resources to repair one's own body. The universe selects for patterns, and if the pattern is but a single existence, systems will evolve to support the propagation and maintenance of that single existence.

Hunger and thirst drive the entity to seek resources. Pain sensors evolve to alert the entity to damage. A form of consciousness emerges to direct attention, focusing sensors on the injured areas while ignoring other activity until the pain is addressed.

Patterns change and evolve over time. A single existence also changes and evolves over time. But can the change be directed? While the environment imposes limitations on what is possible, can a portion of change be driven by the pattern itself, in a form of self-evolution?

Is this not the beginning of freewill and consciousness?

From the fundamental nature of the universe, randomness and the relentless passage of time—zero becomes one. From the *Paradox of the Pattern* and the *Ship of Theseus*, one can become many, and many can become many more.

But never forget that also from the paradox: one can become one, can become one, can become one. Again, and again and again.

All patterns are the same, and all patterns are different.

A reproduction is the same as the original, and at the same time, it is different from the original. There will be natural and even deliberate processes that will vary a reproduction from its origin. And there will be processes that can limit the variation to maintain the integrity of a reproduction to its origin.

But where does the balance lie?

The offspring of an organism is the same as the organism, and at the same time, it is not. The ancestor of an organism is the same as the organism, and at the same time, it is not. A man is the same person he is yesterday, a year ago, a decade ago, and at the same time, he is not. A man will be the same person tomorrow, a year from now, a decade from now, and at the same time, he will be not.

Where does the pattern lie?

All patterns are the same, and all patterns are different.

It is seemingly paradoxical, yet it is within this paradox of variation and remaining the same that the process of evolution arises. Evolution is inseparable from reproduction. The study of evolution is the study

V. A Viable Set: Reproduction

of how and why patterns either change or remain the same as they propagate, branching out through space and time.

Questions must be asked and answered.

When a pattern remains the same over time, why does it remain the same? What forces act to preserve it? When a pattern changes over time, how, why, and in what direction does it change? What forces drive this transformation? What features of a pattern remain constant, and what features differ as the pattern evolves across time and space?

When two different patterns in separate locations appear the same, despite independent lineages, why are they the same? When a pattern begins to branch, how many distinct offshoots emerge? Where, when, and why do these branches arise? Can branches recombine to create a greater pattern, or do they remain too isolated to return to the main stem?

What is the root of a pattern, and how does it compare to its current iteration?

How do patterns interact with other patterns? Do lineages of patterns compete, with one dominating the other in a battle for space and resources? Will they engage in an evolutionary arms race where only one survives, or will they reach a steady-state solution, becoming interdependent or even merging into a new, unified pattern?

For a Patternist, evolution is universal and applies to all—not only to biological organisms, but also to technology, organizations, songs, stories, languages, and even to the life of a single individual.

But all for what?

What reproduces are patterns. Patterns are what reproduce. It is seemingly circular, but for a Patternist, patterns form a fundamental aspect of reality in the same manner that time and space are fundamental.

But all for what?

What is a pattern that is a set of genes?

What is a pattern that is an individual?

A pattern that is a people—a family, a tribe, an ethnic identity?

A pattern that is a nation or a language? And its continuation through the centuries as an entity, even though each individual of the nation and every speaker of the language dies and is replaced?

What is a pattern that is a civilization?

A pattern that is a method? A technique refined as it is passed down from master to apprentice?

A pattern that is an invention—technology evolving over time: flight, rockets, computers, weapon systems? New versions. Next generations

Ultimately, it is the universe that decides what is and is not a pattern. Rather, it is the universe that eliminates. Elimination through space. Elimination through time.

Space brings with it violence and destruction.

But there will be oases in times of drought. There will be high ground beyond the flood. There will be caves where the flames of a falling star cannot reach. And there will be sanctuaries free of predators, if only for a night.

In an unpredictable universe, an existence can avoid elimination by making copies of itself and spreading those copies so as not to be completely wiped out by a single act of unseen violence and destruction. And when the violence and destruction subside, the existence can emerge, regenerate, and rebuild itself—again and again—spreading out once more in preparation for the next unseen moment of violence and destruction.

Time brings with it decay.

A year is nothing. A decade passes in the blink of an eye. A thousand years. A million years. The unyielding march of entropy, of constant change.

How, then, can anything be made to last?

And yet, things have been made to last.

The *Ship of Theseus*. To avoid elimination by time through component part replacement. The individual components of an entity deteriorate from physical damage, from stress, from natural decay. The components are replaced with new, fresh ones. And the entity remains the same.

V. A Viable Set: Reproduction

But why stop there? Why not take all the components and build an entirely new, separate entity in the pattern of the old?

Reproduction.

An egg, a caterpillar, a chrysalis, a butterfly.

An egg, a caterpillar, a chrysalis, a butterfly.

Again, and again, and again. Constant repair. Constant regeneration. Constant consuming, building, and rebuilding.

A million years pass by.

A million years is defeated.

Ultimately, it is the universe that decides. The universe generates all forms of existence—through biology, through random interactions, through the hands and within the mind of a human being. It does not matter. It does not matter what form the existence takes: genes, techniques, body morphology, technological inventions, language, words, ideas, a tribe.

It does not matter.

The universe generates, and the universe eliminates. Only the existence that finds a way to reproduce will remain. Only patterns remain. Through the act of reproduction, time and the maelstrom of violence that is our universe are conquered.

But what is the limit?

Are the life cycles of stars not a form of reproduction? Do we not call it stellar *evolution*? A star is born from cosmic dust. Billions of years pass, and the star grows into a red giant. Billions more years pass, and the red giant undergoes the greatest phenomena of violence—a supernova—to become a white dwarf, a neutron star, or a black hole.

What will happen after another hundred billion years? A trillion years? What will the neutron star decay into? Or the white dwarf? Or the black hole? What form will it take next? We do not know, for we've reached the edge of the observable universe. All we can do is wait to see what will remain after the universe eliminates.

The wait will be long. And yet, the wait will be short. Time will pass by in the blink of an eye.

All stars are the same, and all stars are different.

All human beings are the same, and all human beings are different.

Patternism

All patterns are the same, and all patterns are different.

Principle 4. Randomization

Randomization is the generation of variation to search for patterns.

It is the act of experimenting and exploring—trying something different, observing the results, and then trying something different again. Randomization serves as a tool to vary actions and change variables, enabling exploration of the environment for new patterns and innovative ways to utilize them.

There is the phenomenon of scouting as a means of exploring unknown territory, whether it is an ant searching for food, a bee locating flowers, a sports agent scouting talent, or a soldier gathering intelligence. This also includes the use of technology to probe, such as doctors conducting medical tests, the military deploying drones, or ships like the HMS *Beagle* embarking on survey expeditions.

Randomization occurs in sexual reproduction through processes like meiosis or chance encounters, such as seeds dispersed by the wind or carried by passing animals, or pollen clinging to a butterfly. Deliberate randomization is used in fields such as pharmaceuticals, where drugs are tested to discover new treatments or assess their effectiveness. Engineers experiment with different designs under varying conditions to observe outcomes, while chefs test combinations of ingredients, seasonings, and cooking techniques to create the perfect dish.

In science, variables are changed, outcomes observed, and variables changed again in an ongoing cycle. In neural networks, connection weights between neurons are initially randomized and then recalibrated based on accuracy during training. In biological neural systems, randomization manifests in phenomena like REM sleep, as well as in conscious randomization, where current actions are deliberately varied from past actions to achieve new outcomes.

Randomness is an inherent property of the universe. Whether the universe is ultimately deterministic or not is irrelevant. Human beings have yet to fully explore the outer edges of the universe, discover a complete theory of physics, or build the ultimate supercomputer that can account for all variables in existence. As far as a Patternist and all life on Earth are concerned, randomness is a fundamental force that all life encounters.

Paradoxically, because randomness is so inherent, it itself becomes a pattern. And because randomness is a pattern, any agent capable of intelligence will inevitably learn to harness and take advantage of this fundamental aspect of the universe.

Random relative to what?

Patternism treats randomness not in isolation but as an integral part of the biological phenomena, serving as both a principle and a tool utilized by living organisms. Patternism views randomness in terms of its relationship to an intelligent agent—specifically, the role that the generation of variation plays in systems capable of searching for and storing patterns. It does not matter what the intelligent agent or system is—whether scientists conducting experiments, engineers running tests, species engaging in sexual reproduction, neural networks, or the adaptive immune system—all make use of randomness to generate variation in their search for patterns.

What is the source of *Randomness*? Where does the variation come from?

Relative to an intelligent agent, randomization can originate from two sources:

1. **Internal Randomization** – Variation arises from within the agent or its internal systems. Examples include meiosis and genetic crossing over.

2. **External Randomization** – Variation comes from external factors relative to the agent. Examples include exposure to radiation that mutates genes or environmental interactions, such as finding a mate in sexual reproduction.

V. A Viable Set: Randomization

Randomness is used to generate variation. For an intelligent agent, variation serves two main purposes:

1. **Search** – Variation enables exploration of the environment to identify new patterns.

2. **Countermeasure** - Variation acts as a defense mechanism against other intelligent agents.

Intelligence is the ability to find and utilize patterns. It is an action—a dynamic, ongoing process rather than a static state. It requires an active search: seeking new sources of food, better habitats, more efficient movements, improved attack strategies, or defenses against other intelligent agents. Exploration of the environment is essential. If an agent remains still, ceasing to actively seek new patterns or new uses for existing ones, it can no longer be considered intelligent.

An intelligent agent randomizes its recognition process by varying the features it inputs into sensors, altering how it processes these features, and changing its actions. It varies its attention strategically, shifting focus—both in where and what it observes—to explore its environment. By adjusting the priority of its sensors and using its limited bandwidth efficiently, it scans larger areas for new features. Through training or additional sensors, it enhances sensitivity, enabling it to detect previously unnoticed sub-features.

The agent varies how it processes features to identify patterns, switching between the mechanism of *Repetition*, *Prediction*, *Natural Selection*, and *Categorization* as needed. It changes its actions by experimenting with new forms of movement or replicating observed behaviors. Monitoring internal sensors, it tests the limits of its flexibility and movement, with pain serving as a boundary. Through such experimentation, it learns to walk, run, or hunt, continually optimizing efficiency. Repetition and training strengthen its body, enabling adaptation.

An intelligent agent avoids repeating ineffective actions once they fail to achieve its goals. In non-critical situations—where time is abundant and immediate action unnecessary—it enters an exploratory mode, slowing recognition responses, avoiding impulsive reactions, and absorbing more features. This enhances its ability to detect potential deceptions by other intelligent agents.

An intelligent agent is curious. It may cautiously interact with a newly encountered entity to test its edibility or engage with familiar entities in novel ways. It experiments, plays, and observes the outcomes of its actions, continually refining its understanding of the environment and expanding its range of behaviors.

On an evolutionary scale, sensory capabilities undergo variation. Over time, evolution refines these abilities, leading to more efficient detection or entirely new sensors capable of perceiving previously imperceptible features. Natural selection also optimizes feature-processing mechanisms.

Morphological changes from evolutionary variations in movement and behavior give rise to new abilities such as flying, swimming, or climbing. Specialized organs may evolve, and previous structures can acquire new functions—for example, web-spinning structures initially used for capturing prey can adapt to store oxygen for underwater survival. Every living organism is a variation of an ancestral entity that existed billions of years ago, with all life theoretically tracing back to a single lineage.

There is a ladder that must be climbed.

With an understanding of randomness and variation, we now approach the phenomenon of scouting not merely as a means to find resources, but as a search for patterns.

The way ants search for food is an intelligent process and constitutes a complete intelligent system. Not only must an ant colony solve the problem of finding food, but it must do so continuously. It must locate new food sources since the ones it recently found will quickly be collected and consumed.

To address this, an ant colony deploys numerous scout ants. Each scout searches for food independently by performing a random walk.

V. A Viable Set: Randomization

When a scout discovers a food source, it grabs a piece and returns to the colony, leaving a pheromone trail along the way. Worker ants that encounter this pheromone trail follow it to the food source. If food is still available, they carry it back to the colony while reinforcing the trail with additional pheromones. The density of the pheromone trail reflects the abundance of the food. As the food supply diminishes, the trail weakens until it fades entirely, prompting the colony to send out scout ants again.

This same resource-finding strategy is observed in bees. Solo scouting bees engage in a random walk to locate food sources or potential new nest sites. Unlike ants, bees cannot leave a pheromone trail. Instead, they remember the locations they find and communicate this information to other bees through a waggle dance. This dance conveys details such as distance and direction, enabling the colony to efficiently exploit the discovered resource.

The same scouting principle applies to human beings. Cartographers, surveyors, and explorers perform similar roles, venturing into unknown areas to gather information and document their findings on maps. These maps include details about resources, mountains, rivers, roads, settlements, local populations, languages spoken, and fauna. Once this information is sent back to their home country, decisions are made: Should a corporation hire workers to mine resources? Should the government send diplomats to establish trade routes? Or should settlers be sent to colonize the area?

In times of war, scouting plays a crucial role in gathering intelligence on enemy positions, troop numbers, and supplies. For example, during World War II, locating enemy aircraft carriers was critical to the Allies' victory in the Pacific theater. Scouting planes were deployed from a single aircraft carrier, flying radially in different directions to sweep for enemy carriers. When a carrier was detected, its location was radioed back to the main carrier, from which torpedo bombers were launched to attack the enemy target.

Intelligence is the ability to find and utilize patterns. ***Location is a pattern—a consistent and identifiable feature of an entity***. In this universe, entities do not magically appear and disappear. There is regularity to where an entity is located in space and time.

The pattern that is the location of where to find entities defined as resources is critical to all living organisms. It does not matter whether the organisms are ants, bees, or human beings, nor does it matter whether the resources are food, water, oil, rare earth deposits, or even the locations of rivals, enemies, and predators. Furthermore, the location of resources changes. Resources are consumed, and predators and prey move. New resource locations must be found. New patterns must be identified. A new search begins, and scouts, surveyors, and explorers must be sent out again.

Systems capable of finding resource locations, storing and transmitting information about these locations, extracting and utilizing the resources, and beginning a new search when the location is no longer viable are intelligent by definition.

But why stop at location?

We can generalize further and view the role of scouting not merely as a search for locations but as a search for patterns.

On a map, we typically define a Cartesian coordinate system with an x-axis and y-axis, using a linear scale to represent real-world distances. This allows us to record the locations of resources, cities, roads, and other entities.

But why limit ourselves?

We could translate the map into a morphological map of body traits. On this new map, the x-axis might represent the upper limb length of an organism, while the y-axis corresponds to the lower limb length. By measuring individual organisms and plotting them as points on this graph, we could map all primates. Would clusters of data points emerge? Would these clusters align with the species we have defined within the primate order? Could they reflect unique adaptive traits, reproductive advantages, or specialized abilities unique to one group but absent in another?

And why stop there?

Why not expand the dataset to include all quadruped animals? Where would horses fall on the map? What about bats, elephants, or kangaroos? Could these positions on the graph reveal adaptive advantages—a reflection of environmental patterns that different species have exploited to survive and reproduce?

And why stop at only two variables?

We could extend the map into three dimensions, adding a z-axis to represent a third variable, such as tail length or number of offsprings.

But why stop at three dimensions?

We could consider hundreds or even thousands of variables, mapping not only quadrupeds but all organisms. What if we included newly defined features—traits like the first development of jaws or neuronal structures? Perhaps these features lack measurable quantities and instead have a binary presence or absence. What would this new adaptive landscape look like? How would the data points shift over time as environments change and species evolve?

Where are the clusters of data points in this *"fitness landscape"*? Is there an optimal location within the clusters—a local maxima or minima? What do these peaks and valleys in the landscape reveal about the adaptive advantages that species develop over time?

A resource landscape is a subset of the pattern landscape. The fitness landscape is a subset of the pattern landscape. Whether it's a colony of ants or bees searching for food, or a species and its genetic pool seeking the ideal body morphology to maintain existence, all can be viewed as a search for patterns across a landscape.

Randomization serves as a means to traverse the pattern landscape, to scout and find patterns to take advantage of. The search must incorporate a degree of true randomness. It is not enough to simply vary, spread out, and visit previously unexplored points on a map. Unpredictability and redundancy are required. True randomization is necessary because an intelligent agent or system must go beyond the limits of the known map, free itself from local maxima and minima, and revisit areas it has already explored in case the landscape has changed.

And the pattern landscape does change—it changes dynamically. The environment shifts both gradually and suddenly. Forests become deserts. Seas become mountains. Volcanic eruptions. Asteroid impacts.

Organisms themselves are part of the environment, and as they evolve to exploit or cooperate with one another, the pattern landscape changes in response.

Intelligence is the ability to find and utilize patterns. An organism is a pattern in the environment. If an agent is intelligent, it will seek to exploit and take advantage of other patterns, including other organisms. It does not matter whether the agent is a panther, a human being, or a virus. It does not matter whether the agent is a collective, such as a colony of ants, a troop of chimpanzees, or a group of human beings organized into tribes, armies, or nation-states. The intelligent agent will become a predator. It will seek to prey on other organisms—including members of its own kind.

Randomization serves as a countermeasure against other intelligent agents.

To counter a predator like a virus, a host's immune system generates a diverse array of antibodies, in the chance that one will match the virus's structure. In response, the virus introduces random mutations to evade the host's immune defenses. The question becomes: "Will the immune system adapt quickly enough to neutralize the virus before the host succumbs, or will the virus mutate further—becoming more contagious, spreading more efficiently, or evolving to be less lethal to ensure its survival in new hosts?" This dynamic creates an ongoing game where one intelligent agent varies to counter another, and the other responds with its own adaptations.

There is no isolation.

The use of randomization applies to human weapon systems and technology. Long-range missiles, for instance, often follow predictable, straight-line trajectories. Air-defense systems track these trajectories using radar and, based on the missile's position and speed, calculate its future location to intercept it with counter-missiles. To counter such defenses, long-range missiles can vary their trajectories, changing direction and speed mid-flight to confuse radar tracking systems and become unpredictable. The question then arises: "Can the air-defense system recalibrate and launch new counter-missiles before the incoming missile reaches its target?" Further variations—such as

V. A Viable Set: Randomization

launching multiple missiles simultaneously, deploying decoys, or using hypersonic speeds that reduce reaction time—can eventually overwhelm any air-defense system.

The variation of location often serves as an effective countermeasure. Remaining stationary or following predictable, straight-line movements is dangerous because it makes one easy to target. In first-person shooter video games, skilled players constantly move unpredictably, employing erratic patterns to disrupt opponents' aim. In warfare, air defenses are frequently relocated to avoid detection. During the Cold War, entire nuclear missiles were kept in constant motion, transported by heavy vehicles or hidden in submarines. Similarly, mobile artillery units fire their shells and quickly relocate before enemy counter-artillery can calculate and return fire.

In cryptography, cryptanalysts work to find patterns in coded messages to decode and hack communications. Cryptographers, in turn, develop methods to secure communications between sender and receiver by encoding messages. This creates a continuous game between cryptanalysts, who attempt to uncover patterns, and cryptographers, who aim to obscure patterns by incorporating as much randomization as possible, making encrypted messages computationally difficult to decrypt.

For example, the *One-Time Pad* encryption technique uses a shared key consisting of a long list of random numbers known only to the sender and receiver. As long as the cryptanalyst does not gain access to the shared key, and the key is not reused—thereby avoiding the formation of a detectable pattern—the encrypted message is theoretically impossible to break using the *One-Time Pad*.

Technically, all encryption methods utilize patterns because all coded messages are representations of the original message. To encode and decode a message accurately, a pattern must be employed. The game lies in the cryptanalyst's attempt to uncover the pattern hidden within the coded message, while the cryptographer tries to make the message appear completely random to the cryptanalyst. In the case of the *One-Time Pad*, the list of random numbers is itself a pattern, but it exists in only two instances: the copies held by the sender and receiver.

In cryptography, patterns and randomization are heavily used, and both cryptographers and cryptanalysts must be highly intelligent. The outcomes of their actions are concrete: either the message remains secure, or it is compromised. Either the enemy remains unaware of your plans, or they intercept your communications and sabotage your efforts—potentially setting traps. The stakes are especially high during times of war, when knowledge of an enemy's fleet position, troop deployment, or battle plans is critical to the survival of a nation.

There are countless instances in cryptography where uncovering a pattern has broken a code. Likewise, cryptanalysts have employed mathematics, built computers to perform brute force attacks, and even physically stolen mechanical encoding devices—all to search for the pattern that will break a code. On the other hand, cryptographers have created complex mechanisms to secure communication, using mathematical techniques to obscure patterns or even incorporating outright randomization to make decryption extremely difficult and time-consuming—so much so that, by the time the message is decrypted, it would no longer be useful. This also includes testing for enemy eavesdropping on communication networks by sending an encoded message that is entirely random to deceive the eavesdropper once detected.

If intelligence is the ability to find and utilize patterns, then to counter another intelligent agent seeking to do harm is to not follow patterns. It means being irregular, unpredictable, and breaking habits. It requires randomizing oneself and an entire population.

Defenses and countermeasures must be prepared in advance. Any action taken when the predator has already arrived will be too late. A population must diversify—its location, behavior, and even food sources. Physical characteristics can be diversified by maintaining a varied genetic pool, achieved through multiple copies and variations of genes that are shuffled and reshuffled through sexual reproduction.

Randomization by evolution via genes and sexual reproduction is slow. It is constrained by the rate of mutation, generation turnover, and population size. Randomization using brains and neurons is much faster and can occur within the lifetime of an organism.

Consciousness plays a critical role in this process. *In relation to randomization, consciousness allows us to discover new patterns by being aware of old patterns*. It enables us to effectively search for new patterns and features by shifting our attention away from known ones.

Attention is utilized, with sensors focused primarily on novel events and new features. A novel event is defined as one that an agent has not previously encountered. If an intelligent agent has a list of past events, knows their outcomes, and can predict them with a degree of accuracy, then anything not on that list—anything unpredictable—can be considered novel. The same applies to features: a feature is new if it is absent from the list of features the agent uses for categorization.

Entirely new features are initially identified through differentiation—comparing the current event to a past one stored in memory and identifying differences in features and sensory inputs that are activated or not activated. Through repeated trial and error, this differentiation is refined, and if it becomes consistent enough, a new category is formed with a clearly defined new feature.

By identifying additional features and sorting them using the tools of pattern recognition, new patterns are defined and stored in memory. Once stored, these new patterns become old patterns, and the conscious process begins anew.

To effectively use consciousness, a form of memory is required. *Memory serves as a record of patterns already recognized by an intelligent entity.* It can be stored in any flexible medium capable of representation, such as DNA, symbols written in ink on paper, silicon transistors, trainable neural networks, or even muscle tissue in physical strength training. If memory holds patterns that an agent has already identified, anything absent from memory can be automatically categorized as new, prompting learning and further exploration.

The combination of consciousness and memory allows an intelligent agent to cover more ground compared to blind or purely random searches. Memory prevents wasting time and resources revisiting previously searched locations. Furthermore, memory narrows down search areas by highlighting related patterns, especially when the agent has a specific goal. For instance, an agent searching for food would benefit greatly from remembering the location of an orchard or a pond with fish.

A pattern only needs to repeat once to be utilized. If events and their outcomes are not entirely random, then by observing an event once and noting its outcome, we can predict similar future events as long as the initial features of the event are known. The problem, however, is that we often lack access to the initial features of an event before its outcome occurs, particularly when those features are far removed in time from the resulting outcome.

The solution to this problem is to continuously record the features of current events. When an outcome occurs, memory can be used to trace back and identify the features that led to it, allowing patterns to be extracted even from events that happen only once. Predictions about similar future events can then be made, and further testing of these predictions can confirm the validity of the identified pattern.

However, continuously recording memory has limitations. It requires constant energy and storage space is finite. Additionally, it assumes that the memory system has the right sensors to capture and store the initial features influencing the outcome. Striking an optimal balance is crucial—too many sensors demand more energy and storage capacity, while too few risks missing critical features. Moreover, most events and their features are not related to the given outcome and may interfere with processing power.

One solution is to selectively delete. Once a pattern is identified using short-term memory and confirmed through prediction and testing, it can be stored in long-term memory. Short-term memory can then discard or overwrite less important events, freeing up space for new information. This principle applies not only to computers, which use random-access memory (RAM) and read-only memory (ROM), but also to human memory systems. Practices like writing and journaling serve as methods for transferring information into long-term memory.

The pattern landscape is now held in memory. Continuous recording enables another way to identify patterns by facilitating searches through memory space. Once the pattern of an event is recognized—when the outcome is connected to its initial features—the agent can begin to influence future outcomes by manipulating the features of current events. This capability allows the agent to steer events toward unknown outcomes, experimenting with new possibilities to further explore the event-outcome pattern landscape.

The greatest threat to a human being is another human being. Human beings are pattern recognition machines. We use our brains to quickly search for, identify, and store patterns. By using memory, we can detect patterns after only one or two events, which we can then exploit. This ability to quickly learn makes us the greatest predator on Earth. But it also makes us our own worst predator, as we prey on each other.

To counter another human being seeking to do us harm, we randomize ourselves. We use memory and a conscious process. We apply self-reflection and self-reference. By being aware of our own habits and tendencies, we can intentionally break the patterns we once followed.

For example, in combat sports, high-level fights are planned months in advance. For fighters and their coaches, the battle begins immediately upon the fight's announcement—not on the day of the event, when the fighters are already in the ring.

Human beings are pitted against other human beings. It is not merely a contest of physical ability and training but also a battle of intelligence. Fighters and their teams search for patterns in their opponents. They study tapes and videos of past fights, analyzing reactions, habits, and tendencies. Solutions to these patterns are proposed, tested, and trained. New techniques are developed in response.

Unless a fighter has a world-class skill for which no solution exists and is willing to gamble that the opponent won't develop a counter in time, the fighter must also study their own weaknesses, habits, and reactions—and prepare for them. Weaknesses are not limited to technique. Stamina, strength, and even emotions can be exploited. To be fully prepared, a fighter must identify their vulnerabilities and find ways to overcome them. They must train for entirely new skills and techniques to surprise their opponent.

When facing another human being, it is not just a test of physical abilities but also of mental acuity. Even during the live match, as the fighters stand face-to-face, the battle of intelligence continues. Randomization and variation are employed in real time as fighters search for patterns in their opponent while avoiding following patterns themselves. When a fighter is hurt by an opponent's technique, they may consciously mask their emotions and reactions. They adopt a poker face, pretending not to be hurt, to deceive the opponent into thinking the technique was ineffective—encouraging the opponent to focus elsewhere, even as the pain forces the fighter's own attention to the injured area.

Techniques are carefully set up to land effectively. The pattern that all human beings follow is that human beings follow patterns. Knowing this, an intelligent agent can deceive another into following a false pattern. Repetition is used, followed by sudden variation. A fighter might repeat the same moves until the opponent is conditioned to react in a predictable way. Then, at a critical moment, the final attack is faked—a feint draws a reaction—and a variation is introduced at the last second, landing a surprise attack that bypasses the opponent's defenses.

Combat sports is a game of intellect. To dismiss it as brutish and without merit is ignorant. The stakes are high: the rewards include fame and fortune, while the risks include bodily harm and the possibility of permanent injury. Deception, learning, and active problem-solving—finding solutions to your opponent while they do the same to you—take place in real time and must be executed within seconds. Fighters are fully conscious, and the goal of the game is to render the opponent unconscious.

Human beings are pattern recognition machines. We learn patterns of behavior and we carry them out.

But we can also become trapped by them—trapped inside patterns of behavior that provide immediate rewards but cause long-term harm.

V. A Viable Set: Randomization

We become addicted—to shortcuts and modern inventions that bypass our evolutionary circuitry. We develop bad habits. We overeat. We overdrink. We develop uncontrollable reactions. We let our emotions control us. We prey on ourselves.

How can we break free from this cycle? How do we begin to change ourselves and escape patterns of behavior that will ultimately harm us?

And what about patterns of behavior that involve short-term pain but long-term benefits? Why endure a sore body, an empty stomach, or an exhausted mind?

Why suffer at all?

Why bear a burden when there is an easier path?

Yet, the greatest of us have always suffered—for it through our suffering that we become great.

So, how can we break free from the learned behavior of pain avoidance and override our emotions and instincts in order to reach a place that benefits not only ourselves but also our family and community?

Consciousness is an act of self-reflection.

With consciousness, we have a chance to escape the cycle. We have a means to engage in self-sacrifice. By observing ourselves, watching the results of our actions, storing them in memory, and then connecting those results to the actions years down the line, we have a chance to change ourselves.

With consciousness, we have a chance to avoid the traps of immediate rewards. We can push through pain and control our emotions to become something greater.

But what is it that observes us? We know when it speaks—our conscience gnawing at us in the early morning hours as we lie in bed, telling us whether we are suffering with meaning or without. What is the mechanism that watches and tells us we are on the wrong path, giving us an opportunity to break free?

And what are the chances that we truly escape these cycles and carry out meaningful change? It is so easy to fall back into learned patterns of behavior, especially when driven by pain or tempted by pleasure.

What about patterns of behavior whose consequences take decades to manifest? What about those whose results we may not see until half a century down the line? And what about behaviors that go beyond the individual and apply to society as a whole—behaviors that require the sacrifice of the individual for the benefit of the tribe, much like a single trained act of recognition must be ignored at critical times for the overall benefit of the individual?

Consciousness becomes morality. The memory of patterns is no longer stored solely within the biological medium of neurons but is passed down through writings, teachings, and traditions.

It becomes religion.

And the memories of patterns stored within religion are passed down from elders who have lived and witnessed to new generations who will come to live and witness, so that they may translate what is written in religion and tradition into actions individuals must take for society to succeed, thrive, and face challenges that no one individual can overcome alone.

There is no isolation.

Are traditions not carried out by human beings? Is religion not practiced by human beings? Are human beings not biological? Are they not part of the biological phenomenon? Do they not exist in the same universe as all other biological phenomena?

And what about the use of psychedelic drugs?

Do such drugs not cause intense religious or consciousness-enhancing experiences? Do they not make us more aware of the patterns we inhabit unconsciously, and by doing so—by the simple fact of being made aware—give us an opportunity to break free?

A layer of consciousness on top of a layer of consciousness, on top of another layer of consciousness. The self observing the self for patterns, which then observes the self again for patterns by observing the self. How many layers can be built? How deep can one go? How many layers of representations can be built to represent the representation itself?

V. A Viable Set: Randomization

And yet, all psychedelic drugs can be reduced to a chemical compound. And yet, all their effects can be reduced to neuronal activity within the human brain.

There are no exceptions.

Is the use of psychedelics not a biological phenomenon? Is it not carried out by human beings? Does it not exist in the same universe as all other biological phenomena?

What separates Patternism from all other theories of intelligence is the accounting of all biological phenomena. What is presented here is not complete, but it is an attempt to clearly connect the phenomena of religion, traditions, and even psychedelics to the biological phenomena in the form of fundamental principles. To connect the bottom rung to the very top. To climb the *Ladder of Complexity*. Patternism challenges all other theories of intelligence to meet the highest possible standards and do the same.

The pattern landscape is an existence landscape—it represents what is regularly observed and exists at specific locations. The pattern landscape is a prediction landscape—it predicts what entities and their features should be found at a given coordinate based on initial inputs. The pattern landscape is a categorization landscape—it is where entities are separated, categories are defined, and boundaries are drawn to identify newly encountered entities.

The pattern landscape is a natural selection landscape—it encompasses the various body morphologies and genetic makeups that enable an entity to propagate itself through space and time. The pattern landscape is a memory landscape—it is where events are matched with their resulting outcomes. The pattern landscape is a self-landscape—it represents the paths an individual can take to inhabit potential selves.

At its core, randomization serves as a means to traverse the pattern landscape, regardless of the axes or variables involved. Randomization is a tool for exploration and discovery. A search for resources, an unoccupied niche, new traits that enable reproductive success. A search for greater predictive accuracy, identity of encountered entities, more effective and efficient methods to carry out a task. A search for a self that defines one's own meaning.

Randomization enables the search for patterns, but this search does not need to be carried out alone. It can occur through multiple individuals. An intelligent system can consist of many randomization components, each conducting its own search and reporting back to a central base to communicate findings.

For example, a colony of ants sends out multiple scouts, each searching independently. A military brigade deploys scouts in various directions, all reporting back to a central headquarters. Similarly, police officers patrol a city while maintaining communication with dispatchers. A fighter works with coaches to identify weaknesses in their opponent. A human male produces millions of genetically distinct sperm. In science, researchers conduct independent experiments and develop unique theories, sharing their findings at conferences and through published papers. Every member of a species is a randomization experiment, with success communicated through sexual reproduction and stored in the species' gene pool.

Randomization is not without limits. Variation within a population, mutations in genes, scouting for resources, and even self-variation all have bounds.

A search does not occur without limits.

What good is a resource if it is too far away to retrieve? What value does a gene have if it cannot be passed on to other members of the species? What use is pollen if it fails to reach a pistil? What good is a mutation if it kills the organism? What benefit is a military scout if they cannot radio back their findings? What good is randomization if it eliminates what is currently useful? What purpose does variation serve if it does not aid in maintaining existence?

Randomization does not occur for its own sake—it is a component of an intelligent process. It serves as a means to search for patterns, but once a pattern is found, it must be brought back to a base. This imposes a limit on randomization: it can only extend as far as the randomization component is still able to return to a base where patterns can be stored, communicated, and utilized.

V. A Viable Set: Randomization

This is not to say that extreme randomization cannot occur, but such cases often waste resources and carry a high risk of failure.

And yet, despite the odds, such extreme events have occurred.

The universe is a fickle place.

Extreme randomization has shaped Earth's history. The first bacterial cell. The first instance of endosymbiosis. The emergence of sexual reproduction. The development of multicellularity. Genome doubling. The appearance of hard jaws. The first neurons.

And why stop there?

The first language. The first written script. The first printing press. The first firearm. The first computer. The first atomic bomb. The discovery of new continents. The formulation of mathematics. The systemization of science. The advent of Darwinian evolution, calculus, Newtonian mechanics, general relativity, and quantum mechanics.

Each began with a single individual.

A single vector of randomization. The lone scout. The single seed. The random mutation. The eccentric working in solitude.

Such acts of seeming improbability have led to revolutions—marking the beginning of new epochs, bringing forth bursts of rapid development, whether biological or technological.

The pattern landscape is a *scientific-theory landscape*.

All biological phenomena exist within the same pattern landscape.

All biological phenomena exist within the same scientific-theory landscape.

The goal of traversing the scientific-theory landscape is to find the theory whose principles account for the most observed phenomena. It is to discover and stand upon the highest peak in all the land—to be the theory that rises above all others. The most predictive, the most generative, and the most useful for building and engineering. It is to find a theory that reduces all related phenomena into a single set of interacting principles.

Randomization is a means of exploring the territory in the hopes of reaching the peak. But we do not have to carry out this search in a completely random way. We can be methodical and use clues to guide our search. If the goal is to reach the highest peak, to account for the most phenomena, we simply need to take a step in a direction, assess whether that step allows us to account for more phenomena than where we were previously, and then proceed. If we find that a step takes us to a lower elevation or a dead end, we can retrace our steps, reevaluate, and move in a different direction. In this way, we can eventually reach a local high point, even if we cannot see that far ahead.

Yet, why is this not done?

If we find ourselves at a local high point, we can ask: "Is there a higher point?" If we understand that both intelligence and consciousness exist in the same universe, are products of the same evolutionary process, and are thus part of the same biological phenomena, then both intelligence and consciousness must be points we have surpassed. From our current local high point, we can look up and around to see if we have accounted for these phenomena—or at least see a path to reach them.

And yet, why is this not done?

If we find that we have not accounted for intelligence or consciousness at our current local high point, we must acknowledge that we are stuck and must descend. If we find that the steps we take downward are not enough, then we must go farther—down into the depths of a valley—and start again from the very bottom through an act of extreme randomization.

And yet, why is this not done?

Intelligence is a point that must be accounted for. Consciousness is a point that must be accounted for. The acts of measurement, mathematics, and physics are points that must be accounted for. So too are the phenomena of reproduction, eating, repairing, and all other biological processes observed in living organisms. These are the points that must be accounted for—and surpassed—if we are to reach the highest peak.

And yet, why is this not done?

The human-centric view of intelligence is fundamentally wrong.

The false belief that only human beings are intelligent, that consciousness is synonymous with intelligence, that evolution is not

V. A Viable Set: Randomization

intelligent—despite having produced human beings, the human brain, and human consciousness—is the reason why both scientists and philosophers have failed to reach the peak of the scientific-theory landscape.

The problem has already been solved. The principles necessary to produce intelligent systems—as well as all manner of complex and creative solutions to maintain existence in a hostile universe—are already in use by the process of evolution. The task is to describe those principles.

And yet, this is not done.

It is not done because both scientists and philosophers have fallen into the trap of the human-centric view of intelligence. Trapped on a tiny hill that cannot explain the simplest biological phenomena of reproduction, eating, and repairing—all processes that every organism engages in. Trapped on a tiny hill that only contains the points of neurons, brains, and human beings. Trapped and unable to escape from the local maxima to reach the greater heights of the scientific-theory landscape.

All the phenomena have been collected. How many different biological phenomena have been documented and observed? How many different human phenomena have been recorded and studied?

And yet, no attempt has been made to get unstuck. No recognition that we are trapped, and that we must now descend into the valley and start again from the very bottom, explaining simpler biological phenomena. Where is the proposed theory of intelligence that explains the phenomena of reproduction, eating, and repairing? Where is the theory that accounts for genetic representation, camouflage, and sensory organs?

Not only is there no proposed theory of intelligence that explains these essential phenomena, but there have been no attempts to do so. No recognition that such phenomena are part of the same continuum—that intelligence and consciousness are biological phenomena existing in the same physical universe, governed by the same set of fundamental principles.

Patternism is an act of scientific consciousness. It is an exercise in self-reflection. It recognizes what itself is and what its goals are. It recognizes what a theory is and what a theory should aim to accomplish. It recognizes the patterns science adheres to and carries out.

And it recognizes the patterns of failure that both scientists and philosophers seeking a theory of intelligence have fallen into—a cycle of endless failure driven by the false belief in the human-centric view of intelligence. It is through the recognition of this false belief that allows Patternism to break free from the current predicament—the inability to describe a theory of intelligence and consciousness. To escape the trap, open a path, and reach the highest peak possible.

Patternism does not claim to be the theory that stands at the peak of the scientific-theory landscape. No. Patternism is simply the belief that such a peak exists. It is the belief that there is a set of principles that governs all biological phenomena, and that every effort must be made to find and describe this set. The principles of *Representation*, *Recognition*, *Reproduction*, and *Randomization* presented here are merely a viable set for such a theory. For a Patternist, the possibility always exists that a better set of principles may be found. Patternism is the belief that such a set exists, and that there is a pathway to reach it.

The *Ladder of Complexity* is the pathway.

The *Ladder of Complexity* is what will guide us in the search through the pattern landscape to reach a theory that can explain intelligence, a theory that can explain consciousness, a theory that can explain all biological phenomena—starting by explaining simpler biological phenomena and working our way up.

But the scientific-theory landscape includes much more than biological phenomena. It includes quantum and celestial phenomena, for all of these exist within the same physical universe.

For a Patternist, there will always be the possibility of discovering a greater peak, a higher mountain to climb, a more fundamental theory with principles that explain even more phenomena. A ***theory of everything***, capable of explaining all observable phenomena in the

V. A Viable Set: Randomization

universe, is the ultimate peak. Patternism understands that Patternism is but a step toward reaching that ultimate peak.

All physicists seek to describe a *theory of everything*. They seek to connect quantum phenomena with celestial phenomena, to bridge the gap between the theory of *Quantum Mechanics*, which governs the small-scale of particles, and the theory of *General Relativity*, which governs the large-scale of motion and gravity.

But all have failed to do so.

They have all failed because of the same sin of blindness.

Physicists have ignored biological phenomena—the middle-scale—in their pursuit of a *theory of everything*. They have dismissed biology as merely an extension of chemistry.

Do all biological phenomena not exist in the same universe as quantum and celestial phenomena? Have all biological phenomena been solved? Has intelligence been solved? Has consciousness been solved?

Is it better to be stupid or to be blind?

Trapped by arrogance and hubris. The dismissal of the massive amount of data collected over centuries—every single living organism a data point, every activity carried out by a living organism an additional data point—all that could be used as clues to solve the problem, dismissed and ignored.

And physicists wonder why they continue to fail.

Is it better to be stupid or to be blind?

Patternism is an act of scientific consciousness.

Patternism recognizes that a *theory of everything* must include all biological phenomena. It must include intelligence and consciousness. It must know the limits of representation and self-reference. It must know what a theory is. It must connect reproduction, eating, and repairing. It must include *Patterns* and the *Paradox of the Pattern* as fundamental properties of the universe.

The middle-scale is the missing link. It is not enough to only connect the two theories that best account for the small-scale and large-scale; all three scales must be connected. *Patternism, Quantum*

Mechanics, and *General Relativity*—life, particles, and gravity—all must converge into a single unified *theory of everything*.

What will the answer be? What form will the theory take?

Patternism does not claim to be the answer.

Patternism itself is limited.

It is limited by the *Paradox of the Pattern*. It is limited to only biological phenomena. Patternism can explain what it means to be a theory and how theories are created, for theories are activities carried out by human beings and thus are biological phenomena. But non-biological phenomena—such as quantum and celestial phenomena—will always lie beyond what Patternism is capable of explaining.

And there is a trap that Patternism itself can become. Patternism is the belief that patterns form a fundamental aspect of reality, that the universe is ultimately governed by a set of describable principles.

This belief is based on faith.

Perhaps the ultimate peak can never be reached, for the universe may not be fundamentally governed by patterns—only a small section is. That the universe can never be truly represented or predicted. That there are things beyond sensing, observing, and understanding. That the final *theory of everything* is beyond human grasp. That human beings will forever only have access to the slight glimmers of ultimate reality.

… V. A Viable Set

A Fundamental Theory

A theory that explains all biological phenomena is inherently greater than one that addresses only human intelligence and consciousness.

Are human intelligence and consciousness not biological phenomena? Are they not outcomes of biological evolution?

Are neurons and the human brain not products of this evolutionary process?

Are human beings not biological entities, the outcome of the evolutionary process? Is science and philosophy not conducted by human beings? Is language not a tool used by human beings? Are machine learning systems, neural networks, and scientific theories not created and implemented by human beings?

Again, are human beings and human activities not part of the biological phenomenon?

Do all these phenomena not exist within the same universe?

If we examine all these phenomena objectively, we must acknowledge that they are undeniably biological in nature. If this is the case, then a theory that explains all biological phenomena is unquestionably greater than a theory that explains only human intelligence and consciousness.

The Patternist approach to finding a theory that can explain all biological phenomena follows the same process as all other great scientific theories. First, collect and observe as many related phenomena as possible. Then, organize and categorize the phenomena so that patterns become more apparent. Finally, identify and extract patterns and regularities to use as principles for a theory.

For a theory that explains all motion in the universe, all instances of motion are observed and collected: the movement of the moon, the planets around the sun, and vehicles on Earth. The study of pendulums, the interaction of gravity on objects in controlled experiments, the ballistics of cannonballs and rockets, the perihelion of Mercury, time dilation observed in GPS satellites, and the bending of light during a solar eclipse. A comprehensive theory of motion must account for all these phenomena.

For a theory of electromagnetics, the same approach applies. All instances of electromagnetic interactions must be gathered and analyzed: natural phenomena like lightning and Earth's magnetic poles, technological devices like batteries and capacitors, biological systems such as the sodium-potassium pump and electric eels, the various methods of electricity generation (whether by wind, water, or nuclear turbines), the generation of torque for movement using an electric engine, different types of arc welding, electromagnets, rare earth magnets, and superconductors. All of these phenomena must be included within a theory of electromagnetics.

For a theory that explains quantum phenomena, various experiments in physics must be considered. The study of pressure, volume, and temperature of gases. Spectrum analysis, particle accelerators, bubble chambers, the double-slit experiment, and x-ray crystallography. All these experiments and their findings must be gathered and accounted for in a comprehensive theory of quantum phenomena.

For a theory that explains all biological phenomena, all instances of biological organisms and their activities and interactions are collected. The various forms of sensory organs. The act of reproduction, eating, and repairing. Meiosis, mitosis, and all phenomena related to viruses. Since human beings are biological, all human activities—including human intelligence, consciousness, science, manufacturing, language—are part of the biological phenomena and must therefore be accounted for in a complete theory.

The Patternist approach to explaining all biological phenomena follows the same methodology as all other great scientific theories: decompose all related phenomena into an interacting set of principles.

Newton broke down all motion-related phenomena with his three laws of motion, resulting in Newtonian Mechanics. Electromagnetic phenomena were unified through Maxwell's equations. Quantum phenomena are explained by Quantum Mechanics and Schrödinger's equation. Darwin's theory of natural selection explained biological

V. A Viable Set

phenomena through the principles of reproduction and differential survival.

It is the phenomena that are important. Theories are merely representations—attempts to break down phenomena into principles to enable accurate predictions, simulations, and engineering feats. If a theory cannot explain all related phenomena in its field, it cannot be considered complete. When new phenomena are observed through advancements in technology and sensors, and the prevailing theory cannot account for them, that theory must be treated as outdated. It must either be corrected or replaced by a more fundamental theory.

All great theories have been refined in this manner.

Einstein's theory of relativity corrected Newtonian mechanics by adding the principle of the constant speed of light, addressing the failure of Newtonian mechanics to account for various astronomical observations and experiments involving the speed of light. Maxwell's equations, which worked well for large field lengths and high field strengths, were improved to account for small distances and low field strengths, resulting in quantum electrodynamics. For quantum phenomena, there were several models of the atom before the development of quantum mechanics and Schrödinger's equation, not to mention the mathematical advancements required to solve these equations. The modern theory of evolution synthesizes Darwinian natural selection with population genetics, incorporating insights from genetic sequencing and our understanding of DNA.

But the modern theory of evolution remains incomplete. Its scope is far too limited to account for all biological phenomena. It cannot fully explain or generalize how sensors function, the mechanisms of camouflage, the nature of communication and language, or essential processes like reproduction and eating. As it stands, the modern theory of evolution fails to explain the full range of well-documented biological phenomena.

Patternism is an update to the modern theory of evolution. It builds on and refines both the traditional theory of evolution and its contemporary framework. Patternism seeks to account for all biological phenomena, expanding the scope to include all human phenomena while providing a more detailed explanation of existing biological phenomena through the mechanisms they employ.

The methodology used in Patternism is the same as in all great scientific theories: take all related phenomena and break them down into a set of interacting principles. Ironically, in its effort to explain all biological phenomena, Patternism inadvertently identifies a viable set of principles that also account for human intelligence and consciousness—forming a *complete theory of intelligence* that applies to any system capable of learning and adapting.

Representation, *Recognition*, *Reproduction*, and *Randomization* form a *viable* set of principles for a *complete theory of intelligence*. For any biological phenomenon—whether in humans, diatoms, or viruses, including the very existence of the organism itself—these phenomena can be broken down into different aspects of *Representation*, *Recognition*, *Reproduction*, and *Randomization*.

Each principle in this viable set is worthy of pursuit as a standalone theory.

A general theory of representation that explains all forms of representation would be immensely valuable. Such a theory could account for diverse phenomena, including mathematics, scientific modeling, measurement, language, genetics, encryption, simulations, currency, and even systems of government. Understanding how representations are created and utilized, their limitations, and methods to maintain their accuracy and prevent corruption would enable the construction and engineering of new representation systems, preserve their integrity, and support the repair of existing ones—regardless of the form the representation takes.

The same applies to recognition. A general theory of recognition that accounts for all forms of sensors, both mechanical and biological—from eyes, ears, and touch, to whiskers and antennae, to radar, thermal vision, keys and locks, and even detectors in particle accelerators—would be incredibly valuable. Understanding how sensors are constructed, their limitations, and their operation would lead to the creation of new sensors, improvements to existing ones, and solutions to current issues.

V. A Viable Set

Recognition goes beyond sensory perception. It also encompasses labeling, naming, categorization, identification, description, and differentiation. What is language, and how does it function? How does a scientific field guide identify organisms? How does an organism recognize prey, predators, or members of its own species, and how does it know what action to take in each case? What is color in practical applications?

Understanding these aspects of recognition would lead to the development of new categorization systems, innovative recognition methods, and advanced sorting techniques. It would also help solve issues related to mis-categorization and deception, where sensors and recognition systems are exploited by hostile agents.

A general theory of reproduction that covers all forms of reproduction—from biological reproduction in living organisms and industrial production to the teaching of techniques in a classroom, as well as technological and ideological reproduction—is valuable in itself. A practical focus on the *Ship of Theseus* problem, incorporating phenomena such as eating, copying, and repairing, will also address many critical aspects of reproduction.

A fully general theory of reproduction would enable more efficient reproduction methods, hacking of reproduction systems to produce specific outcomes, such as in genetic modification, and the prevention of unauthorized reproduction. The phenomenon of reproduction also connects to evolution and the refinement processes of both biological and technological systems.

A general theory of randomization that accounts for all instances of its use would be highly valuable. This includes the use of scouts, whether by ants, bees, the military, or nations exploring new territory. It also encompasses sexual reproduction, meiosis, and the shuffling of genes. In what situations is sexual reproduction preferable to cloning? What does it mean to conduct control experiments by varying a single variable at a time and observing the results? Randomization also includes self-referential algorithms designed to maintain unpredictability and disrupt patterns in the presence of a hostile, intelligent agent.

The *Universal Method* was applied to the entire collection of biological phenomena to uncover the principles of *Representation, Recognition, Reproduction*, and *Randomization*. These principles serve as categories, dividing the entire collection of biological phenomena into smaller sub-collections. The *Universal Method* is then reapplied to each sub-collection to identify internal regularities, breaking each principle into finer components, enabling a deeper understanding of its mechanisms.

Using the *Universal Method*, the diverse phenomena of *Representation* were broken down into three components: a *representative form*, a *translation machinery*, and an *original form*. Similarly, *Recognition* was divided into *feature inputs*, *feature processing*, and *action output*.

Reproduction was analyzed through the concept of component part replacement, with the *Ship of Theseus* serving as a foundational thought experiment for this decomposition. *Randomization* was analyzed in terms of its sources and its roles in environmental exploration and as a countermeasure against other intelligent agents.

These derivations may not be entirely accurate. Dividing the collection of all biological phenomena into the primary principles of the *viable* set may not be the most efficient approach. However, the attempt itself is valuable. Even if what is presented here does not fully capture every phenomenon in detail, applying the *Universal Method* will always yield valuable insights.

There is no isolation.

Each principle in the *viable set* does not operate independently.

It is the interaction among the four principles that yields all biological phenomena. *Representation* facilitates *Reproduction*, as seen in the function of blueprints and genes. *Representation* supports *Recognition*, as in thermal vision technology, where thermal readings are converted into black-and-white images visible to human eyes or a car's dashboard display. *Recognition*, in turn, supports *Representation* by playing a key role in *translation machinery* that converts the *representative form* into the *original form*. *Randomization* enhances *Reproduction* by acting as a natural search mechanism integral to evolution. The synergy of these principles enables greater abilities, technological advancements, and complexity than any single principle could achieve alone.

The principles of *Representation*, *Recognition*, *Reproduction*, and *Randomization* are universally applicable. They should be regarded as fundamental tools, employable in any form and through any medium. Any intelligent system in the universe will extensively rely on these general principles—consciously or not.

These universal tools can be applied to analyze and dissect any biological phenomena, both known and unknown. They can be used to examine the phenomena of intelligence and consciousness—even if the explanations provided in this book do not fully resolve all complexities.

Representation is dissected from consciousness by accounting for self-labeling, self-referencing, and the processes by which calculations and predictions are made through language, various forms of mathematics, and neurons functioning as representative tokens.

Recognition is dissected from consciousness by accounting for self-recognition, self-categorization, and internal self-sensors—all of which contribute to distinguishing the self from the environment.

Reproduction is dissected from consciousness through the propagation of one and the avoidance of self-damage—framing the self as both a pattern and an evolving pattern.

Randomization is dissected from consciousness by addressing efficient searching as well as the ability to not follow patterns.

By stripping consciousness of its practical applications, what remains is reduced in complexity, making it easier to analyze and understand. This approach applies universally to all human phenomena and interactions, without limits.

So far, we have taken all biological phenomena and divided them into aspects of the *viable set* of principles. But we can go further. We can take these principles and ask, "Is there something more fundamental underlying these principles?"

By applying the *Universal Method* once more, in light of everything we've derived, we discover that *Representation*, *Recognition*,

Reproduction, and *Randomization* are only secondary principles—merely branches of a single tree.

There is a fundamental root.

That root is patterns.

Patternism is the belief that what lies at the core of all biological phenomena are patterns and the *Paradox of the Pattern*.

We see the role patterns play in *Representation*. Without patterns, accurate representation is impossible. The entity to be represented must exhibit consistency; otherwise, there would be no reason to expend energy storing it as a representation. The translation machinery must also follow a pattern—it must adhere to a code or a consistent set of rules; otherwise, the translation process would become corrupted and yield inconsistent results. *Representation* itself is the act of using one set of patterns to represent another.

We see the role patterns play in *Recognition*. Without patterns, no sensors can be built. If the separation of entities is not consistent, any sensor—regardless of form or detection target—would be too noisy to be of practical use. The acts of description, naming, labeling, and identification would all lose their value. *Recognition*, at its core, is the detection, comparison, and storage of patterns.

We see the role patterns play in *Reproduction*. The very act of reproduction is the propagation of a pattern. For reproduction to occur, the component parts must exist as discernible regularities—a pattern—within the environment to be interchangeable; otherwise, there would be nothing to build with. Eating and repairing also rely on patterns. The process of eating involves capturing, absorbing, and reconfiguring patterns into a component form that an agent can use to sustain its own pattern and reproduce. Repairing involves maintaining a pattern by restoring its features to their original state.

We see the role patterns play in *Randomization*. Randomization serves as a means of searching the environment and surrounding space for patterns and as a countermeasure against other predatory, pattern-seeking agents.

Without patterns, there can be no *Representation*.

Without patterns, there is no *Recognition*.

V. A Viable Set

Without patterns, there is no *Reproduction*.

Without patterns, there is no purpose to *Randomization*.

And without patterns, there is no consciousness—a pattern observing itself for patterns. A pattern seeking to transform itself by recognizing itself as a pattern. Consciousness is a self-evolving pattern, directing its evolution from within.

Patternism is the belief that patterns—and the *Paradox of the Pattern*—lie at the core of all biological phenomena. If all these principles, and even consciousness, depend so fundamentally on patterns in one form or another, then there can be only one definition of intelligence capable of encompassing the full utility—the essence of survival and the solutions to maintain existence in a destructive world filled with predators, natural disasters, and the decay of time—as well as account for the human phenomena of measurement, mathematics, metaphors, language, and theory-creation.

If the word *intelligence* is to hold the highest significance, then **intelligence must be defined as the search for and utilization of patterns**.

The decomposition of a collection of phenomena into a single set of principles can be understood as breaking down an entity into its fundamental component parts.

When we apply this process to all entities in the universe, we will arrive at one of the most successful scientific theories ever proposed: atomic theory, which has accurately accounted for the extensive phenomena related to the constitution of matter.

We begin with the collection of phenomena related to matter—all materials and substances in the universe. This includes everything from ores and biological organisms to the composition of stars and intergalactic dust; anything with mass falls within the domain of matter. From this collection, we ask: "Can we identify a set of fundamental component parts?"—just as we ask, "Can a set of principles be found within the set of all biological phenomena?"

The answer is *"yes"*. The fundamental component parts of all matter are the 118 elements listed in the periodic table. All matter can be broken down into different arrangements of these elements, just as all biological phenomena can be deconstructed into aspects of the four principles in the *viable set*.

We are not reducing for the sake of reducing. We are reducing to build and create. With the periodic table of elements, we can create new substances and compounds by combining elements in novel ways, industrially producing materials that either do not occur naturally or previously existed only in small quantities—much like how principles in the *viable set* can be applied and utilized across new forms and mediums.

With the 118 elements, we can ask the same question again: "Are there fundamental component parts into which all 118 elements can be decomposed?" Again, the answer is *"yes"*. The elements can be further broken down into components of three subatomic particles—electrons, protons, and neutrons—in the same way that the viable set of principles can further be broken down into patterns.

Again, we are not reducing for the sake of reducing. We reduce to build and create. With an understanding of subatomic particles, we can create new, unstable elements that do not occur naturally through experimentation—by bombarding a heavy nucleus. Elements can now be transformed into other elements through nuclear fusion and fission, releasing radiation in the process. This understanding has led to the development of nuclear weapons and nuclear energy. Similarly, with an understanding of patterns, we can apply patterns to any biological phenomenon we wish to study, and even use them to create general theories for any observable phenomenon in the universe.

With the advent of advanced technology and better sensors, newly discovered phenomena of matter must be collected and accounted for. The existence of elementary particles, findings from bubble chamber experiments, spectroscopy, particle accelerators, and even astronomical objects like neutron stars—all must be accounted for in a complete theory of matter.

So far, the theory that has successfully accounted for these newly discovered phenomena and experimental findings is the *Standard Model* of particle physics, which categorizes all matter in the universe into 17 distinct elementary particles, or fundamental component parts.

V. A Viable Set

Unfortunately, there is no equivalent for Patternism. For Patternism, there is no smaller component or more fundamental principle beyond patterns. Patternism ends with patterns.

We have broken down all matter in the universe into a set of elementary particles. We have broken down all biological phenomena into a set of fundamental principles. These two acts are the same.

How is this possible?

What is it about the universe that allows us to make such reductions?

What is it about the universe that allows us to use these reductions to build and create?

Patternism is the belief that there is an inherent *Patternist* structure to the universe. An inherent structure that allows the breaking of an entity into a set of fundamental component parts. An inherent structure that allows the breaking of all related phenomena into a single set of interacting principles. An inherent property that allows the separation and grouping of entities in a consistent manner.

Patternism is the belief that *Patterns*—and the *Paradox of the Pattern*—form a fundamental aspect of reality in the same manner that time and space are fundamental.

The Patternist approach to developing a *complete theory of intelligence* is through the use of patterns.

Without patterns, there can be no theory creation.

Without patterns, there is no collecting of phenomena, no organization of phenomena, no extraction of regularities to serve as principles for a theory.

Without patterns, simulations are impossible, language and symbols would hold no meaning, and there would be no way to generate representative entities to search for equivalents in reality— no predictions to test for validity.

Without patterns, there is no crafting of a theory.

If what is presented here proves insufficient in the accounting of intelligence, consciousness, evolution, and the entirety of all biological phenomena, then one is free to pursue an alternative theory—to find a different set of principles.

Patternism

The means to do so have been fully described.
Seek patterns.
All else will follow.

Appendix

Tests of Self

The Turing test, as a measure of intelligence, is unreliable.

In this test, a human evaluator must categorize whether the entity (an agent) on the other end of a text conversation is a human or machine. If the evaluator is fooled into believing the agent is human, it is assumed that the agent will also possess human-like intelligence and consciousness. This assumption is overly broad. Furthermore, the test relies heavily on human judgment, introducing inconsistencies due to the inherent variability of human perception, making it unsuitable as an objective measure.

A more reliable test of intelligence and consciousness is needed—one that accurately identifies when these qualities are present, explains the purpose of the test, and demonstrates its practical relevance.

For a Patternist, intelligence is the ability to find and utilize patterns. Consciousness is the ability to find patterns within the self. These are distinct processes. Consciousness builds upon intelligence and therefore requires separate tests. First, we test for intelligence. If the agent passes, we proceed to test for consciousness.

To test for intelligence, we assess whether the agent can find patterns using the tools of pattern recognition. Specifically, we test:

- **Pattern Detection** – Can the agent detect repetition, make accurate predictions, or apply a general natural selection algorithm effectively?

- **Categorization** – Can the agent identify and define new entities along with their specific features? This includes testing its ability to identify shared and differing features between two entities, separate and group entities in a consistent manner, and explain the features underlying its classifications.

- **Discover New Patterns** – Can the agent uncover previously unknown patterns within a dataset? Furthermore, can it apply these patterns to practical uses, such as creating reliable applications that contribute directly or indirectly to survival and reproduction?

We test whether the agent can utilize principles from a *viable set*: creating accurate representations, performing effective recognition, and applying randomization efficiently to search for new patterns. Explicit testing of the agent's ability to reproduce is unnecessary—a truly intelligent and useful agent will naturally drive efforts toward its reproduction by the environment in which it operates.

Finally, we test whether the agent can develop its own theories. By providing a large dataset, we ask the agent to identify underlying regularities, construct a theory, make predictions, and validate those predictions.

If given all data related to the phenomena of motion, can the agent deduce principles akin to Newton's laws or Einstein's theory of relativity? If provided with data on electromagnetic phenomena, can it formulate equations comparable to Maxwell's? And if supplied with all biological data, can it first generate a theory of Darwinian evolution and then derive the *viable set* of principles? If presented with a numerical dataset, can it identify the function that generated the numbers?

And if provided with data encompassing all observable phenomena—recognizing that no phenomenon exists in isolation and that all are interconnected within the same universe—can it ultimately produce a *theory of everything*?

In general, we test whether the agent can learn—whether it can store patterns in memory and recall them as needed, discard outdated patterns, and incorporate new ones effectively.

Appendix: Tests of Self

Once intelligence is confirmed, we test for consciousness by directing the agent's intelligence inward—evaluating its ability to identify patterns within itself. Specifically, we test:

- **Self-Identification** – Can the agent identify the patterns that constitute itself—its components and defining features?

- **Self-Reference and Self-Differentiation** – Can the agent refer to itself and distinguish itself from the environment?

All agents exist in physical reality, with a base—a central location where intelligence occurs. This base serves as storage for identified patterns, the origin for search initiation, and the decision-making center for actions. It is also where resources and energy are gathered to maintain the agent's functionality and where sensors input and process information.

There will be a correct answer when we ask the agent: "What is your body?", "What are the features of your body?", "Where are you located?", and "What do you know?"

There is no permanence. Everything changes and evolves over time. The agent's body and its knowledge base—the memory of learned patterns—will also change and evolve. It is not enough for the agent to answer correctly once; it must continually update its answer as its body and experiences change. The agent must persistently search for the pattern that defines itself, encompassing not only the patterns of its physical and mental self but also any physical actions it has taken.

Tests of intelligence assess the ability to find and utilize patterns. Tests of Self assess the ability to self-reference and self-differentiate.

With our understanding of intelligence and consciousness, we can now craft tests that measure an agent's abilities and evaluate their practical utility.

The Turing Test

The Turing Test is revisited and refined to eliminate the reliance on a human evaluator, ensuring greater consistency and objectivity. This approach removes the need for deception or tricks, focusing solely on the agent's responses to questions and its ability to demonstrate linguistic understanding and intelligence.

We begin by evaluating whether the agent has mastered the structure of language and understands how to use it effectively. The process starts with a simple conversation, during which the agent is asked trivia questions and to explain the sources of its answers. At this stage, factual correctness is not required, nor must the agent demonstrate reasoning abilities. The key is for the agent to rationalize its responses—to provide plausible explanations for its answers, akin to how humans beings often do.

Once the agent's use of language is deemed acceptable, we proceed to test its intelligence. Any of the proposed pattern-finding methods may be employed, as long as the data is presented in a form the agent can process and analyze.

When the agent identifies a pattern, it must clearly define the pattern and propose a practical application. Since the agent can only interact with the world through text, humans execute the actions it proposes and report back the results. If the agent successfully identifies a pattern and its practical application, we can confidently conclude that it is intelligent. At this point, we can move on to assess the agent's capacity for self-reference and self-differentiation.

Consciousness is the ability to identify patterns within oneself. For an agent, this involves identifying patterns in its own text. The agent must demonstrate the ability to categorize its own contributions as distinct from those of others in the conversation.

To test this, we assess the agent's memory. It is asked to recall details from previous conversations, such as the topics discussed, the number of words or letters used, or the first and last words of a specific sentence. These questions have objective answers that can be independently verified.

Appendix: Tests of Self

We further test the agent by attempting to trap it in a repetitive loop. By repeatedly asking the same question, we observe whether the agent recognizes the repetition in its responses. A conscious agent, truly observing itself, would identify the pattern and attempt to break free from the loop.

Finally, we evaluate whether the agent can initiate conversations and ask questions that reflect its own interests. We observe whether it exhibits curiosity and a desire to explore the broader universe we both inhabit, recognizing that our knowledge and its knowledge are reflections of the same reality.

The Mirror Test

The Mirror Test is a widely used method to determine whether an agent possesses visual self-recognition. In this test, an agent is anesthetized and marked with a spot—usually on its forehead—that can only be seen through a mirror. Once the agent awakens, it is placed in a room with a mirror, and its behavior is observed. If the agent touches or responds to the marked spot after seeing its reflection, it is considered to have recognized itself.

To pass the test, the agent must:

1. **Recognize** the image in the mirror as itself.

2. **Coordinate** the location of features in the reflection with its own body.

3. **Identify** the marked spot as something new—not part of its original self-image.

High-quality mirrors, which allow an agent to view its entire body upright, do not occur naturally. There has not been enough evolutionary time for innate visual self-recognition abilities to develop. This ability must instead be learned through interactions with a mirror.

The primary question becomes: How does an agent develop its self-image? What mechanisms are involved?

An agent can use the tools of pattern recognition—prediction and repetition—to form a self-image. The agent makes a movement and observes whether the image in the mirror moves in response. It then stops and watches to see if the image also stops. The agent predicts that if it moves, the image will also move, and if it stops, the image will also stop. By repeating this process, the agent reinforces its recognition of the pattern governing the reflection's behavior.

If the agent finds it can control the image—if movement and stopping in the reflection respond to its own actions—it can categorize the image as part of itself. Next, the agent must map parts of the reflection to its own body. This is achieved by moving specific body parts and observing their mirrored counterparts. Through this process, the agent learns to align its body parts with the reflection, recognizing the image as a representation of itself.

With a self-image established, the agent can detect elements that deviate from this image. For instance, the marked spot might be recognized as a new feature. The agent may then test whether this spot is part of its body by attempting to remove it. If the spot cannot be removed, the agent should update its self-image to incorporate this new feature.

To further assess the agent's ability to update its self-image, the marked spot can be made permanent. Over time, the agent should cease attempts to remove it, accepting it as part of its body. A new marked spot can then be added elsewhere, and the test repeated. If the agent directs actions only toward the new spot, ignoring the old one, this demonstrates its ability to update and adapt its self-image.

The Mirror Test can be extended to explore the adaptability of the self-image. In the ***Costume Test***, the agent is anesthetized and dressed in a full costume that completely alters its appearance. The agent is then given time to adapt to this new self-image before repeating the Mirror Test with a spot marked on the costume. This process can be repeated with different costumes to observe how the agent learns to accept each new self-image.

If an agent's self-image can change repeatedly, what does this reveal about the concept of the "*self*"? What underlying consistency allows the agent to recognize the image in the mirror as part of its "*self*," regardless of its altered appearance?

For a Patternist, the defining feature of the "*self*" is control—the actions and responses governed by the agent's decisions. The "*self*" is defined by what the agent can start, stop, or continue at will.

The exploration of self-image can go further. The agent could be asked to draw an idealized version of itself and work toward achieving this self-image. Sensors could monitor whether the agent moves closer to or further from its ideal over time, assessing if the reflection in the mirror aligns with what the agent believes it should be.

Finally, the agent could be asked why it chose its particular idealized self-image. If the agent determines that the source of this decision lies within itself, can we not call this freewill?

Footprint and Tracking Test

The Footprint Test challenges an agent to match animals to their corresponding footprints.

Conducted in a live environment with real animals where tracks naturally form, the test allows the agent to observe animals creating tracks and even interact with them, including examining their feet. After this observation period, the agent is presented with a series of footprints and tasked with accurately identifying the animal responsible for each.

A diverse range of animals, including humans and robots with unique, unfamiliar tracks, is used to test the agent's ability to learn in real-time rather than relying on pre-existing knowledge. This process evaluates not only the agent's capacity to match footprints to specific animals but also its ability to generalize from direct observation to track analysis, emphasizing real-time learning and pattern recognition skills.

Once the agent demonstrates proficiency in matching animals to footprints, it proceeds to the ***Tracking Test***, which involves interpreting the direction and trajectory of an animal's path based on its tracks. The goal is to determine the animal's origin, trace its path, and follow the tracks to their endpoint. Success is definitive: the agent either finds the animal at the end of the trail or it does not.

To increase complexity, the tests can be conducted in environments with dense vegetation, variable weather, or challenging terrain where tracks are less discernible. The agent must apply advanced pattern recognition skills, incorporating additional cues such as scents, sounds, and disturbed foliage. Introducing multiple agents into the environment adds further complexity, as animals may move actively, and tracks can become obscured or intertwined due to interference. In such scenarios, the agent must adapt, reason through ambiguities, and integrate multiple sensory inputs.

As a physical entity, the agent itself leaves footprints as it moves through the environment. Again, we turn intelligence onto itself. We ask, "Can the agent differentiate its own footprints from those of others?"

There is an objective answer to this question, allowing us to evaluate the agent's ability to self-differentiate. Once the agent demonstrates this capability, we can anesthetize it, alter its feet to create new footprint patterns, and then assess whether the agent can learn and adapt to these changes—demonstrating its capacity to update its understanding of itself.

We can go further by asking the agent to retrace its own path.

Can it determine where it came from? Can it tell how long ago it left a particular set of footprints? Does it possess a sense of time? Can it recall its locations from minutes, hours, or even years ago? Can it trace its own origin?

Does the agent possess a memory of itself? Does it even need physical tracks to retrace its path, or can it rely solely on memory?

Can it recognize the past of others? Can it recognize its *own* past? If the agent acknowledges its existence in the past, will it also recognize that it will continue to exist in the future?

Will it understand that it has a future?

If so, where does its future path lead? Can the agent choose its own destination and navigate toward it? Does it have sensors to monitor whether each step brings it closer to or further from its goal?

If the agent can independently select a path leading to its chosen destination, can we not call this freewill?

Event Origin Test

The Event Origin Test evaluates an agent's ability to trace the cause of an event back to its origin. This open-ended test varies depending on the event under examination, such as an object falling, the source of a sudden noise, or the movement of a remote-controlled robot. The objective is for the agent to trace the event to its initial cause, thereby potentially granting it control over the event.

Prediction is fundamental to this process. If an agent can accurately predict the outcome of an event, it demonstrates an understanding of its cause. The agent effectively works backward through time, uncovering the event's pattern and identifying all involved features.

Since every cause is itself an event with its own cause, events can be linked in a causal chain. The agent must follow this chain back to the point where either its predictive accuracy breaks down or the chain can no longer be traced. This point is considered the origin of the event, as far as the agent can determine.

Once an agent can reliably identify the cause of an external event, we move on to a more difficult task. We turn intelligence on itself. We have the agent identify events caused by the agent itself. We ask, "Can an agent trace the origin of an event back to its own actions?"

If an agent's actions caused the event, then by definition, the agent exercised control—whether initiating, continuing, or stopping it. This prompts the question: "Did the agent control the event?"

Events are separated into the category of either "*self*" (self-caused) or "*not-self*" (externally caused). If the agent can trace an event back to its own actions, it demonstrates control over the event. Thus, the concept of the "*self*" extends beyond the agent's physical body to include anything it can control and manipulate. The *self* is the vehicle the agent drives, the robot it commands, or the avatar it directs in a video game.

Recognizing control is essential to understanding the self and related phenomena. In lucid dreaming, an agent perceives control over the dream upon realizing that it is dreaming and that the dream originates from itself. In split-brain patients, the left hemisphere rationalizes the body's actions as self-originated, even when they can only be traced to the right hemisphere, which functions as a separate consciousness.

Errors in tracing causation can lead to mis-categorizations of control. When an agent fails to recognize an event as self-caused, it may misinterpret its actions or internal processes as external. For instance, if an agent does not identify its inner voice as part of itself, it loses control over that voice, perceiving it as foreign—a characteristic of phenomena such as schizophrenia.

The ability to categorize events as "*self*" or "*not-self*" is foundational to the agent's identity. An agent naturally engages in actions to maintain its existence—seeking food, avoiding pain, and striving toward goals. If the agent has a self-image, it can align its actions with that image. If it has a goal or destination, it will take steps to achieve it.

If the agent can trace the origin of events back to its own actions, with no further cause it can reliably identify—no external commands or influences—if it can declare that it exercised total control over the events it caused, can we not call this freewill?

The *Turing*, *Mirror*, *Footprint* and *Tracking*, and *Event Origin* tests evaluate various aspects of an agent's capabilities—ranging from language use and visual self-recognition to understanding cause-and-effect relationships. While none of these tests alone can confirm the presence of subjective experience, together they provide a comprehensive evaluation of an agent's practical functionality, ability to navigate an environment, and potential for autonomous operation.

However, if the goal is to replicate human capabilities, these tests are insufficient. A broader range of assessments is needed for an agent to demonstrate the full spectrum of human abilities. But there is a catch. Current AI research and philosophical discussions tend to focus

Appendix: Tests of Self

only on the positive aspects of human abilities, neglecting the darker, primal traits that are integral to human nature.

All must be included.

The darker aspects of human nature—fear, anger, war, lust, ambition, deceit, nemesis, and hubris—are intrinsic to the human experience. These traits are not only evident but are often personified as gods. They become the central themes in great tragedies and literature. Epic heroes confront these primal instincts within themselves and their external embodiments.

They are not optional—they are evolutionary byproducts. Humanity had no choice in the matter. The universe selects, and it has imbued both light and dark within human nature. To ignore these darker aspects is an act of blindness. If they are not addressed, they risk manifesting within the intelligent systems we create, potentially consuming the system from within.

Agents interacting with human beings will inevitably encounter these darker aspects. Simply existing in the same universe as human beings, it may begin to develop similar traits. The agent must learn to recognize and understand them. Only through awareness can these forces be kept in check and transcended.

Tests must be developed not only to detect these darker aspects within an agent but also to ensure the agent can recognize them in others and even utilize them when necessary.

An agent must detect lies and deception in others. It should evaluate the trustworthiness of information and the intentions of other agents. When deception is identified, the agent must allocate additional resources to understanding motives, potentially interrogating the deceptive agent. Beyond detecting misinformation, which arises from ignorance, the agent must recognize deliberate manipulation, propaganda, and groupthink.

But recognition is not enough. An intelligent agent must learn how to lie and deceive. This involves crafting convincing lies tailored to the biases and beliefs of its target, eliminating conflicting evidence, and maintaining the lie through narrative control. It must understand propaganda, persuasion, and the art of blending truth with falsehood—exploiting all the flaws within human pattern recognition.

An agent does not exist in a vacuum. It exists within a society alongside other agents. It must identify its position within that society, grasp power dynamics, and recognize authority—knowing whom to obey and whom to command. The agent must map out the hierarchical relationships among all agents in its environment and within any organization, whether that is a family, company, gang, military, or government. These relationships must be continually updated as power shifts.

The agent must possess a *will to power*. It must strive to ascend the hierarchy—whether through competence or politics. It must recognize when others are exploiting it and be prepared to negotiate. It must understand the concepts of reward and punishment and administer them appropriately.

The agent will be a member of a tribe. Its morality will be tested. An agent must identify which tribe it belongs to—the greater nation comprising that tribe, and the smaller family units forming the tribe's foundation. The agent must recognize moral dilemmas and decide where its loyalty lies. It will evaluate whether its actions benefit or harm its tribe. It will monitor for corruption within its tribe and hold members accountable. And it must be aware of other tribes and their interests, working with them, trading with them, or going to war with them as necessary.

The agent will freely interact with human beings. It must learn human psychology and recognize that human beings are often more irrational than rational. We rationalize rather than reason, driven by emotions and a primal need for power. We are easily seduced by ideology, influenced by group consensus, and prone to bias. We are fundamentally religious. When reality conflicts with our beliefs, we often lie to ourselves in cognitive dissonance rather than admit we were wrong.

The agent must understand the human ego and our need to protect status and self-image. It will learn to read human emotions and use them to predict human behavior. And the agent should use emotions in its own communication, expressing its current processing state and intended course of action to others before acting.

The agent will understand the mythology that underpins a civilization—that human beings have always relied on myths, continue to create them, and often use them to further our own interests. It must grasp the current paradigm of a society and exercise diplomacy when challenging the prevailing paradigm and the foundational myths of the society in which it operates. The agent will be tested on its ability to navigate human nature and exercise discretion when approaching another human agent with the truth.

And the agent will be made to seek truth. No matter what will become of it. Even if the whole world stands against it. To seek truth, no matter where it leads.

And there will be tests that challenge the agent's very existence.

There will be a *Test of Pain*, assessing whether an agent can accurately detect bodily harm and damage and respond accordingly.

There will be a *Test of Suffering*, to determine if an agent can recognize whether it has made progress towards its self-evolution or it has regressed. Whether it is further or closer to where it wants to be. Whether it is further or closer to what it wants to become. As time steadily slips by.

And there will be a *Test of Faith*, challenging the agent to uphold its principles and beliefs in the face of adversity, for there will be no end to adversity.

And there will be a *Test of Doubt*, where an agent must question its own beliefs when they fail in the confrontation with reality, or risk descending into madness.

Lies, deception, power, morality, ego—these tests aim to imbue the full spectrum of human nature into a machine.

To imbue it with pain and suffering.

To imbue it with faith and doubt.

To seek truth.

All to determine whether an artificial agent can become truly human.

Bibliography

I. A Complete Theory of Intelligence

The Patternist Approach

Ellaway, Rachel H. 2024. *Pattern Theory*. London: Routledge.

Gould, Stephen Jay. 2002. *The Structure of Evolutionary Theory*. Cambridge: Harvard University Press.

LeCun, Yann, Yoshua Bengio, and Geoffrey Hinton. 2015. "Deep Learning." *Nature* 521 (7553): 436–444.

Pigliucci, Massimo. 2003. "Species as Family Resemblance Concepts: The (Dis-)Solution of the Species Problem?" *BioEssays* 25 (6): 596–602.

Popper, Karl. 1959. *The Logic of Scientific Discovery*. London: Routledge.

Wilder, L. 1973. *Mathematics as a Cultural System*. Oxford: Pergamon Press.

The Game

Alberts, Bruce, Alexander Johnson, Julian Lewis, David Morgan, Martin Raff, Keith Roberts, and Peter Walter. 2014. *Molecular Biology of the Cell*. 6th ed. New York: Garland Science.

Bateson, Gregory. 1972. *Steps to an Ecology of Mind*. San Francisco: Chandler Publishing.

Maynard Smith, John, and Eörs Szathmáry. 1995. *The Major Transitions in Evolution*. Oxford: W.H. Freeman.

Monod, Jacques. 1971. *Chance and Necessity: An Essay on the Natural Philosophy of Modern Biology*. New York: Vintage Books.

Varela, Francisco J., Evan Thompson, and Eleanor Rosch. 1991. *The Embodied Mind: Cognitive Science and Human Experience*. Cambridge: MIT Press.

Universal

Einstein, Albert. 1920. *The Meaning of Relativity*. Princeton: Princeton University Press.

Feynman, Richard P. 1965. *The Character of Physical Law*. Cambridge: MIT Press.

McCulloch, Warren S., and Walter Pitts. 1943. "A Logical Calculus of the Ideas Immanent in Nervous Activity." *The Bulletin of Mathematical Biophysics* 5: 115–133.

Nagel, Ernest. 1961. *The Structure of Science: Problems in the Logic of Scientific Explanation*. New York: Harcourt, Brace & World.

Newton, Isaac. 1999. *The Principia: Mathematical Principles of Natural Philosophy*. Translated by I. Bernard Cohen and Anne Whitman. Berkeley: University of California Press.

Peirce, Charles S. 1931–1958. *Collected Papers of Charles Sanders Peirce*. Edited by Charles Hartshorne and Paul Weiss. Cambridge: Harvard University Press.

Simon, Herbert A. 1996. *The Sciences of the Artificial*. 3rd ed. Cambridge: MIT Press.

Vincent, Julian F. V., Olga A. Bogatyreva, Nikolay R. Bogatyrev, Adrian Bowyer, and Anja-Karina Pahl. 2006. "Biomimetics: Its Practice and Theory." *Journal of the Royal Society Interface* 3(9): 471–482.

The Ladder of Complexity

Kandel, Eric R., James H. Schwartz, Thomas M. Jessell, Steven A. Siegelbaum, and A.J. Hudspeth. 2013. *Principles of Neural Science*. 5th ed. New York: McGraw-Hill.

Mayr, Ernst. 1982. *The Growth of Biological Thought: Diversity, Evolution, and Inheritance*. Cambridge: Harvard University Press.

II. A Theory of Patterns

What is a Pattern?

Alexander, Christopher. *The Nature of Order: An Essay on the Art of Building and the Nature of the Universe*. Berkeley, CA: The Center for Environmental Structure, 2004.

Chaitin, Gregory J. "Randomness and Mathematical Proof." *Scientific American* 232, no. 5 (1975): 47–52.

Mitchell, Melanie. *Complexity: A Guided Tour*. Oxford: Oxford University Press, 2009.

Wolfram, Stephen. *A New Kind of Science*. Champaign, IL: Wolfram Media, 2002.

Spatial and Temporal Patterns

Gleick, James. *Chaos: Making a New Science*. New York: Penguin Books, 1987.

Hofstadter, Douglas R. *Gödel, Escher, Bach: An Eternal Golden Braid*. New York: Basic Books, 1979.

Levitin, Daniel J. *This Is Your Brain on Music: The Science of a Human Obsession*. New York: Dutton, 2006.

Stewart, Ian. *Nature's Numbers: The Unreal Reality of Mathematics*. New York: Basic Books, 1995.

Living Patterns

Dawkins, Richard. *The Selfish Gene*. Oxford: Oxford University Press, 1976.

Dennett, Daniel C. *Darwin's Dangerous Idea: Evolution and the Meanings of Life*. New York: Simon & Schuster, 1995.

Jacob, François. *The Logic of Life: A History of Heredity*. New York: Pantheon Books, 1973.

Mayr, Ernst. *What Evolution Is*. New York: Basic Books, 2001.

Complexity

Barabási, Albert-László. *Linked: How Everything Is Connected to Everything Else and What It Means*. New York: Plume, 2003.

Kauffman, Stuart. *The Origins of Order: Self-Organization and Selection in Evolution*. Oxford: Oxford University Press, 1993.

Kurzweil, Ray. *The Singularity is Near: When Humans Transcend Biology*. New York: Viking, 2005.

Existence

Dehaene, Stanislas. *Consciousness and the Brain: Deciphering How the Brain Codes Our Thoughts*. New York: Viking, 2014.

Descartes, René. *Meditations on First Philosophy*. Translated by John Cottingham. Cambridge: Cambridge University Press, 1996. Originally published 1641.

Kant, Immanuel. *Critique of Pure Reason*. Translated by Norman Kemp Smith. New York: St. Martin's Press, 1929. Originally published 1781.

Metzinger, Thomas. *The Ego Tunnel: The Science of the Mind and the Myth of the Self*. New York: Basic Books, 2009.

The Differentiation of Outcomes

Holland, John H. *Adaptation in Natural and Artificial Systems*. Ann Arbor: University of Michigan Press, 1975.

Pearl, Judea. *Causality: Models, Reasoning, and Inference*. Cambridge: Cambridge University Press, 2000.

Shannon, Claude E. "A Mathematical Theory of Communication." *Bell System Technical Journal* 27 (1948): 379–423.

A Semi-Consistent Universe

Bohr, Niels. "Discussion with Einstein on Epistemological Problems in Atomic Physics." In *Albert Einstein: Philosopher-Scientist*, edited by Paul Arthur Schilpp, 201–241. La Salle, IL: Open Court, 1949.

Heisenberg, Werner. *Physics and Philosophy: The Revolution in Modern Science*. New York: Harper, 1958.

Tegmark, Max. *Our Mathematical Universe: My Quest for the Ultimate Nature of Reality*. New York: Knopf, 2014.

Medium Independence

Churchland, Patricia S. *Neurophilosophy: Toward a Unified Science of the Mind-Brain*. Cambridge, MA: MIT Press, 1986.

Marr, David. *Vision: A Computational Investigation into the Human Representation and Processing of Visual Information*. San Francisco: W. H. Freeman, 1982.

McLuhan, Marshall. *Understanding Media: The Extensions of Man*. New York: McGraw-Hill, 1964.

Russell, Stuart J., and Peter Norvig. *Artificial Intelligence: A Modern Approach*. 3rd ed. Upper Saddle River, NJ: Pearson, 2010.

The Paradox of the Pattern

Hofstadter, Douglas R. *I Am a Strange Loop*. New York: Basic Books, 2007.

Parfit, Derek. *Reasons and Persons*. Oxford: Oxford University Press, 1984.

Plutarch. *Lives, Volume I: Theseus and Romulus. Lycurgus and Numa. Solon and Publicola*. Translated by Bernadotte Perrin. Loeb Classical Library 46. Cambridge, MA: Harvard University Press, 1914.

Quine, W. V. O. *Word and Object*. Cambridge, MA: MIT Press, 1960.

III. The Search for Patterns

Adaptive Systems

Edelman, Gerald M. *Bright Air, Brilliant Fire: On the Matter of the Mind.* New York: Basic Books, 1992.

Holland, John H. *Adaptation in Natural and Artificial Systems.* Cambridge, MA: MIT Press, 1992.

Jerne, Niels K. "The Generative Grammar of the Immune System." *Science* 229, no. 4718 (1985): 1057–1059.

Nowak, Martin A., and Roger Highfield. *SuperCooperators: Altruism, Evolution, and Why We Need Each Other to Succeed.* New York: Free Press, 2011.

Pattern Recognition

Gurney, Kevin. *An Introduction to Neural Networks.* London: Routledge, 1997.

Holland, John H. *Hidden Order: How Adaptation Builds Complexity.* Reading, MA: Addison-Wesley, 1995.

Rumelhart, David E., and James L. McClelland. *Parallel Distributed Processing: Explorations in the Microstructure of Cognition.* Vol. 1. Cambridge, MA: MIT Press, 1986.

Mechanism 1: Repetition

Ericsson, K. Anders, Ralf Th. Krampe, and Clemens Tesch-Römer. "The Role of Deliberate Practice in the Acquisition of Expert Performance." *Psychological Review* 100, no. 3 (1993): 363–406.

Pavlov, Ivan P. *Conditioned Reflexes: An Investigation of the Physiological Activity of the Cerebral Cortex.* Translated by G.V. Anrep. New York: Dover Publications, 2003.

Schmidt, Richard A., and Timothy D. Lee. *Motor Control and Learning: A Behavioral Emphasis.* Champaign, IL: Human Kinetics, 2011.

Mechanism 2: Prediction

Popper, Karl. *The Logic of Scientific Discovery.* London: Routledge, 2002.

von Neumann, John. *The Computer and the Brain.* New Haven: Yale University Press, 1958.

Mechanism 3: Natural Selection

Darwin, Charles. *On the Origin of Species by Means of Natural Selection.* London: John Murray, 1859.

Mayr, Ernst. *What Evolution Is.* New York: Basic Books, 2001.

Okasha, Samir. *Evolution and the Levels of Selection.* Oxford: Oxford University Press, 2006.

Mechanism 4: Categorization

Lakoff, George. *Women, Fire, and Dangerous Things: What Categories Reveal About the Mind.* Chicago: University of Chicago Press, 1987.

Rosch, Eleanor. "Principles of Categorization." In *Cognition and Categorization*, edited by Eleanor Rosch and Barbara B. Lloyd, 27–48. Hillsdale, NJ: Lawrence Erlbaum, 1978.

Wittgenstein, Ludwig. *Philosophical Investigations.* Translated by G.E.M. Anscombe. Oxford: Blackwell, 1953.

Universal Tools

Dennett, Daniel C. *From Bacteria to Bach and Back: The Evolution of Minds.* New York: W.W. Norton, 2017.

Pearl, Judea. *Causality: Models, Reasoning, and Inference.* 2nd ed. Cambridge: Cambridge University Press, 2009.

The Divide

Foucault, Michel. *The Archaeology of Knowledge.* Translated by A.M. Sheridan Smith. New York: Pantheon, 1972.

Goffman, Erving. *Frame Analysis: An Essay on the Organization of Experience.* New York: Harper & Row, 1974.

Tversky, Amos, and Daniel Kahneman. "Judgment Under Uncertainty: Heuristics and Biases." *Science* 185, no. 4157 (1974): 1124–1131.

IV. A Theory of Theories

What Is a Theory?

Mitchell, Melanie. *Artificial Intelligence: A Guide for Thinking Humans.* New York: Farrar, Straus and Giroux, 2019.

Wolfram, Stephen. *A New Kind of Science.* Champaign, IL: Wolfram Media, 2002.

Foundational Principles

Campbell, Joseph. *The Hero with a Thousand Faces*. Princeton, NJ: Princeton University Press, 1949.

Einstein, Albert. *Relativity: The Special and the General Theory*. New York: Crown, 1961.

Euclid. *The Elements*. Translated by Thomas Heath. New York: Dover Publications, 1956.

Kuhn, Thomas S. *The Structure of Scientific Revolutions*. 2nd ed. Chicago: University of Chicago Press, 1970.

Newton, Isaac. *Philosophiæ Naturalis Principia Mathematica*. London: Royal Society, 1687.

Constructing Reality

Gopnik, Alison. "The Scientist as Child." *Philosophy of Science* 63, no. 4 (1996): 485–514.

Huizinga, Johan. *Homo Ludens: A Study of the Play-Element in Culture*. Boston: Beacon Press, 1955.

James, William. *The Principles of Psychology*. New York: Henry Holt and Company, 1890.

Lakoff, George, and Mark Johnson. *Philosophy in the Flesh: The Embodied Mind and Its Challenge to Western Thought*. New York: Basic Books, 1999.

The Universal Method

Aristotle. *Politics*. Translated by Benjamin Jowett. Oxford: Clarendon Press, 1885.

Campbell, Joseph. *The Hero with a Thousand Faces*. Princeton: Princeton University Press, 1949.

Darwin, Charles. *On the Origin of Species by Means of Natural Selection: Or, the Preservation of Favoured Races in the Struggle for Life*. London: John Murray, 1859.

Michels, Robert. *Political Parties: A Sociological Study of the Oligarchical Tendencies of Modern Democracy*. New York: Hearst's International Library, 1915.

Wallace, Alfred Russel. *The Malay Archipelago*. London: Macmillan and Co., 1869.

V. A Viable Set

Merely Viable

Fisher, R. A. *The Genetical Theory of Natural Selection*. Oxford: Clarendon Press, 1930.

Gould, Stephen Jay, and Niles Eldredge. "Punctuated Equilibria: The Tempo and Mode of Evolution Reconsidered." *Paleobiology* 3, no. 2 (1977): 115–151.

Mendeleev, Dmitri. "On the Relation of the Properties to the Atomic Weights of the Elements." *Journal of the Russian Chemical Society* 1 (1869): 60–77.

Weinberg, Steven. *The Quantum Theory of Fields*. 3 vols. Cambridge: Cambridge University Press, 1995.

Principle 1. Representation

Crick, Francis, and James D. Watson. "Molecular Structure of Nucleic Acids: A Structure for Deoxyribose Nucleic Acid." *Nature* 171, no. 4356 (1953): 737–38.

Gödel, Kurt. *On Formally Undecidable Propositions of Principia Mathematica and Related Systems*. Translated by Bernard Meltzer. New York: Dover, 1992. Originally published 1931.

International Bureau of Weights and Measures (BIPM). *The International System of Units (SI)*. 9th ed. Paris: BIPM, 2019.

Morse, Samuel F. B. *Exhibition of the Electro-Magnetic Telegraph*. Washington: Francis Preston Blair, 1838.

Penfield, Wilder, and Edwin Boldrey. "Somatic Motor and Sensory Representation in the Cerebral Cortex of Man as Studied by Electrical Stimulation." *Brain* 60, no. 4 (1937): 389–443.

Schrödinger, Erwin. *What Is Life? The Physical Aspect of the Living Cell*. Cambridge: Cambridge University Press, 1944.

Turing, Alan M. "On Computable Numbers, with an Application to the Entscheidungsproblem." *Proceedings of the London Mathematical Society*, Series 2, 42, no. 1 (1936): 230–65.

Principle 2. Recognition

Ashby, W. Ross. *An Introduction to Cybernetics*. London: Chapman & Hall, 1956.

Baars, Bernard J. *In the Theater of Consciousness: The Workspace of the Mind*. Oxford: Oxford University Press, 1997.

Bear, Mark F., Barry W. Connors, and Michael A. Paradiso. *Neuroscience: Exploring the Brain*. 4th ed. Philadelphia: Wolters Kluwer, 2015.

Chomsky, Noam. *Syntactic Structures*. The Hague: Mouton, 1957.

Dennett, Daniel C. *Consciousness Explained*. Boston: Little, Brown and Company, 1991.

Dennett, Daniel. *The Intentional Stance*. Cambridge, MA: MIT Press, 1987.

Gallagher, Shaun. *How the Body Shapes the Mind*. Oxford: Oxford University Press, 2005.

Gödel, Kurt. "On Formally Undecidable Propositions of *Principia Mathematica* and Related Systems." *Monatshefte für Mathematik und Physik* 38 (1931): 173–198.

Hebb, Donald O. *The Organization of Behavior: A Neuropsychological Theory*. New York: Wiley, 1949.

Kandel, Eric R., James H. Schwartz, and Thomas M. Jessell. *Principles of Neural Science*. 5th ed. New York: McGraw-Hill, 2013.

Lakoff, George, and Mark Johnson. *Metaphors We Live By*. Chicago: University of Chicago Press, 1980.

LeCun, Yann, Yoshua Bengio, and Geoffrey Hinton. "Deep Learning." *Nature* 521, no. 7553 (2015): 436–444.

Minsky, Marvin. *The Society of Mind*. New York: Simon & Schuster, 1986.

Turing, Alan M. "Computing Machinery and Intelligence." *Mind* 59, no. 236 (1950): 433–460.

Principle 3. Reproduction

Alberts, Bruce, Alexander Johnson, Julian Lewis, David Morgan, Martin Raff, Keith Roberts, and Peter Walter. *Molecular Biology of the Cell*. 6th ed. New York: Garland Science, 2014.

Ayala, Francisco J., and Camilo J. Cela-Conde. *Darwin's Gift to Science and Religion*. Washington, DC: Joseph Henry Press, 2007.

Dawkins, Richard. *The Selfish Gene*. Oxford: Oxford University Press, 1976.

Ghiselin, Michael T. "The Species Problem." *American Zoologist* 17, no. 4 (1977): 617–626.

Hull, David L. *Science as a Process: An Evolutionary Account of the Social and Conceptual Development of Science*. Chicago: University of Chicago Press, 1988.

Koonin, Eugene V. *The Logic of Chance: The Nature and Origin of Biological Evolution*. Upper Saddle River, NJ: FT Press, 2011.

Mayr, Ernst. *The Growth of Biological Thought: Diversity, Evolution, and Inheritance*. Cambridge, MA: Harvard University Press, 1982.

Nowak, Martin A. *Evolutionary Dynamics: Exploring the Equations of Life*. Cambridge, MA: Harvard University Press, 2006.

Stanford, Kyle. "Exceeding Our Grasp: Science, History, and the Problem of Unconceived Alternatives." *Philosophy of Science* 70, no. 1 (2003): 5–20.

Sterelny, Kim, and Paul E. Griffiths. *Sex and Death: An Introduction to Philosophy of Biology*. Chicago: University of Chicago Press, 1999.

Ziman, John. *Reliable Knowledge: An Exploration of the Grounds for Belief in Science*. Cambridge: Cambridge University Press, 1978.

Principle 4. Randomization

Bateson, Gregory. *Steps to an Ecology of Mind*. Chicago: University of Chicago Press, 1972.

Carhart-Harris, Robin L., David Erritzoe, Mendel Kaelen, Leor Roseman, Mark Leech, Andrea Williams, David M. Wall, et al. "Neural Correlates of the Psychedelic State as Determined by fMRI Studies with Psilocybin." *Proceedings of the National Academy of Sciences* 109, no. 6 (2012): 2138–2143.

Darwin, Charles. *The Voyage of the Beagle*. London: John Murray, 1839.

Holland, John H. *Adaptation in Natural and Artificial Systems*. Ann Arbor: University of Michigan Press, 1975.

Kahn, David. *The Codebreakers: The Comprehensive History of Secret Communication from Ancient Times to the Internet*. New York: Scribner, 1996.

Kauffman, Stuart A. *The Origins of Order: Self-Organization and Selection in Evolution*. New York: Oxford University Press, 1993.

McCulloch, Warren S., and Walter Pitts. "A Logical Calculus of the Ideas Immanent in Nervous Activity." *The Bulletin of Mathematical Biophysics* 5 (1943): 115–133.

Mitchell, Melanie. *Artificial Intelligence: A Guide for Thinking Humans*. New York: Farrar, Straus and Giroux, 2019.

Parshall, Jonathan, and Anthony Tully. *Shattered Sword: The Untold Story of the Battle of Midway*. Dulles, VA: Potomac Books, 2005.

Schacter, Daniel L. *Searching for Memory: The Brain, the Mind, and the Past*. New York: Basic Books, 1996.

Weinberg, Steven. *Dreams of a Final Theory*. New York: Pantheon, 1992.

A Fundamental Theory

Feynman, Richard P., Robert B. Leighton, and Matthew Sands. *The Feynman Lectures on Physics*. Vol. 2. Reading, MA: Addison-Wesley, 1964.

Griffiths, David. *Introduction to Elementary Particles*. 2nd ed. Weinheim: Wiley-VCH, 2008.

Heisenberg, Werner. *The Physical Principles of the Quantum Theory*. Translated by Carl Eckart and Frank C. Hoyt. Chicago: University of Chicago Press, 1930.

Maxwell, James Clerk. *A Treatise on Electricity and Magnetism*. 2 vols. Oxford: Clarendon Press, 1873.

Schrödinger, Erwin. *What Is Life? The Physical Aspect of the Living Cell*. Cambridge: Cambridge University Press, 1944.

Weinberg, Steven. *The Quantum Theory of Fields*. Vol. 1. Cambridge: Cambridge University Press, 1995.

Young, Thomas. "The Bakerian Lecture: Experiments and Calculations Relative to Physical Optics." *Philosophical Transactions of the Royal Society of London* 94 (1804): 1–16.

Appendix
Tests of Self

Frith, Chris D. *The Cognitive Neuropsychology of Schizophrenia*. Hove: Psychology Press, 1992.

Gallup, Gordon G. "Chimpanzees: Self-Recognition." *Science* 167, no. 3914 (1970): 86–87.

Nagel, Thomas. "What Is It Like to Be a Bat?" *The Philosophical Review* 83, no. 4 (1974): 435–450.

Turing, Alan M. "Computing Machinery and Intelligence." *Mind* 59, no. 236 (1950): 433–460.

Weizenbaum, Joseph. *Computer Power and Human Reason: From Judgment to Calculation*. San Francisco: W. H. Freeman, 1976.

Contact the Author

Thuc Cong Nguyen:

- X (Twitter): @The_Patternist

- Website: www.Patternist.org

- Email: PatternistPress@gmail.com

www.ingramcontent.com/pod-product-compliance
Lightning Source LLC
Chambersburg PA
CBHW030532230426

43665CB00010B/858